Metal-to-Metal
Adhesive Bonding

Metal-to-Metal Adhesive Bonding

STEFAN SEMERDJIEV Dipl.Ing.

BUSINESS BOOKS LIMITED

LONDON

First published 1970

© STEFAN SEMERDJIEV 1970

All rights reserved. Except for normal review purposes,
no part of this book may be reproduced or utilized in any form
or by any means, electronic or mechanical,
including photocopying, recording, or by any information
storage and retrieval system, without permission of the publishers

ISBN 0 220 79950 4

This book has been set 11 on 14 pt Times and printed by
C. Tinling & Co. Ltd, Prescot, Lancs.
for the publishers, Business Books Limited
(registered office: 180 Fleet Street, London EC4)
publishing offices: Mercury House, Waterloo Road, London SE1

MADE AND PRINTED IN GREAT BRITAIN

Contents

	Preface	vii
1	**Introduction**	1
	1.1 Welded and soldered joints	
	1.2 Riveted and bolted joints	
	1.3 Press joints	
	1.4 Adhesive bonded joints	
2	**Basic principles of adhesion**	7
3	**Adhesives—composition, classification, properties**	12
	3.1 Composition of adhesives	
	3.2 Classification of adhesives	
	3.3 Structural adhesives	
4	**Processes for adhesive bonding**	
	4.1 Surface preparation	
	4.2 Preparing and applying the adhesive	
	4.3 Joining the parts to be bonded and curing the adhesive	
	4.4 Additional processing	
5	**Mechanical properties and performance of adhesive bonds**	55
	5.1 Bond strength	
	5.2 Stress distribution in bonded joints	
	5.3 Static strength of bonded metal-to-metal joints	
	5.4 Dynamic strength of bonded metal-to-metal joints	
	5.5 Ageing and resistance to environment conditions	
	5.6 Some possibilities for increasing the strength of bonded metal-to-metal joints	
6	**Testing and inspection of adhesive bonds**	78
	6.1 Destructive methods for testing bonded metal joints	
	6.2 Non-destructive inspection methods for adhesive bonds	

CONTENTS

7 Design of adhesive bonded joints   97
    7.1 Joints used in adhesive bonding
    7.2 Dimensioning adhesive bonded joints
    7.3 Combining riveted, bolted and welded joints with adhesives for metals

8 Application of adhesive bonding of metals   115
    8.1 Applications
    8.2 Reconditioning and repair with adhesives for metals

9 Bonded sandwich structures   158
    9.1 Materials for sandwich structures
    9.2 Manufacture of honeycomb cores
    9.3 Making honeycomb sandwich structures
    9.4 Design features
    9.5 Application of honeycomb sandwich structures

Appendix 1
Suppliers and manufacturers of specific adhesives   173

Appendix 2
Testing adhesives for metals and adhesive bonds: methods, standards and specifications   176

References and general bibliography   179

Index   189

Preface

Adhesive bonding of metals and other non-porous materials (glass, ceramics, etc.) is of exceptional interest in all spheres of engineering and industry. In addition to fabrication advantages, this new method of forming load-bearing joints offers the best utilization of the strength of the bonded components, while at the same time, providing a stable configuration.

Designers have found that bonding gives a new approach to many design problems. This is valid not only for the design of the joint, but also for the selection of the materials used. Bonding, moreover, permits certain constructions that could not be achieved otherwise.

The use of bonding often results in savings due to simplified design and production techniques. Adhesive bonding can also offer an improvement in quality and strength, weight savings, fewer component parts, and an increase in work productivity.

Adhesives are not new and bonding in general has gained wide acceptance as a fastening method in everyday practice. But in introducing metal-to-metal bonding the psychological difficulties of overcoming prejudices, fears and hesitation are much greater than the technological difficulties.

It is true that almost all materials can be bonded together if the correct procedure is used and if a favourable geometry for the joint design is selected. It is true that comparatively few engineers are familiar with adhesive bonding and have yet had experience in bonding processes. For this reason, and because of existing prejudice, industry in general has been slow in adopting bonding techniques and in exploiting their full potential. Old prejudices die hard and many engineers still find it difficult to have confidence in metal-to-metal bonds. It is therefore desirable that the problems of adhesive bonding of metals should be explained and made clear.

Adhesives have only been in use for bonding metals for about 25 years, but during this time a considerable amount of experience has been accumulated. It has been estimated that structural adhesives now have an annual, future growth rate of about 20 per cent per year. In an economy growing at a far lower rate, this must mean that these adhesives have advantages over other joining methods that engineers and production people cannot ignore.

This book is concerned mainly with the use of adhesive metal-to-metal bonding in various branches of industry and engineering (mechanical, electrical, civil, etc.). On account of this, consideration is given to the methods of bonding, as well as to properties, design, testing and inspection of bonded joints. The basic principles of adhesion, the chemical composition and preparation of the adhesives, and the curing reactions are dealt with briefly. A table of suppliers and manufacturers of specific adhesives for bonding different materials is included. Many varied applications in which bonding has been used are discussed. Notice should be taken of the fact that some of the techniques mentioned in the book may be covered by patents.

A separate chapter gives greater detail on honeycomb sandwich constructions, the manufacture of which has become possible during the last few years with the development of adhesive bonding. These constructions are capable of very high strength- and rigidity-to-weight ratios and their weight is very low. The strength of honeycomb sandwich constructions can be greater than that of sheet material of the same weight: up to 16 times if it is of steel, and 10 times if it is of aluminium. Because of their excellent properties and high structural efficiency, honeycomb sandwich constructions are coming increasingly into general engineering, having been well tested in the aircraft industry.

As industry in Britain is adopting the metric SI system, SI units or suitable multiples or sub-multiples of these units are used in the book in accordance with the British Standards.

Since the examples given in the book cover many spheres of technology, it may well be that readers, better versed than the author in their special field, will detect errors or omissions. For this reason the author and the Publishers will accept with gratitude any suggestions.

With this book, which is his first attempt in English, the author hopes to help in overcoming the guarded attitude of many engineers towards a process that is relatively new, and thus contribute to the wider application of adhesives in bonding metals.

S.S.

1
Introduction

The chemical industry has undergone a considerable world-wide expansion during the last few decades and as a result of intensive research in the field of organic chemistry many new materials have been developed. Plastics occupy an important place among these materials. Initially, they were thought of as substitutes for materials already in use, and engineers saw them as inferior materials. But practice has shown that, correctly applied, plastics not only replace several widely used materials, but in many cases they even excel them. For example, until not very long ago, adhesives for wood, paper, leather, rubber, ceramics, and other materials were produced only from vegetable, animal or mineral substances. However, advances in synthetic adhesives based on plastics displaced these products. Synthetic products are distinguished above all by a stronger adhesion and a greater resistance.

The good adhesion of some synthetic adhesives to metals was established a few decades ago, but the systematic development of these adhesives, first applied in the aircraft industry, started only during the Second World War.

With correct design and the proper use of a suitable adhesive, reliable bonded metal assemblies of great strength can be produced. In these, failure under stress will occur in the metal and not in the bond. However, in order to obtain good results a careful study of a variety of factors influencing the quality and the properties of bonded joints is needed.

It should not be assumed that adhesive bonding will altogether replace welding, riveting and other mechancial joints now in wide use. Just as the application of these conventional methods for fastening depend in every individual case on the specific conditions, so bonding has marked features of its own. Certain advantages and disadvantages of the various fastening methods are discussed and compared briefly below.

1.1 Welded and soldered joints

Welding is possible only if the parts to be joined are of similar material. The high temperature of welding may cause changes in the microstructure of metals, adversely influencing the strength. Moreover, internal stresses develop during cooling, causing warpage and distortion of the structure. The welding of various metals is linked with great difficulties, and a loss in strength. This applies particularly to high-strength steels and light metal alloys. The weldability of some of them is very poor, and for this reason they are not welded in practice.

Soldering makes the joining of dissimilar metals possible; the danger exists, however, of galvanic corrosion (when joining steel and aluminium, for example). The removal of corrosive fluxes calls for costly neutralizing operations. Furthermore, soldering requires various expensive materials (for example solder is about 30 per cent dearer per kilogram than the most expensive modified epoxide adhesive).

1.2 Riveted and bolted joints

Drilling or punching of matching holes is needed; this weakens the parts to be jointed and adds to the problems of sealing. What is more, a non-uniform distribution of stresses takes place when the joints are loaded, peaks appearing around the holes. This requires the use of thicker materials causing an increase in weight of the structures. Non-uniform stress distribution results also in low fatigue strength. Joining of light gauge materials presents further difficulties. Other disadvantages are: danger of capillary and contact corrosion, unevenness of the surfaces, and loosening caused by cyclic loads. In addition, joining with rivets and bolts is generally a slow and labour-consuming process.

1.3 Press joints

Joining by cold- and warm-pressing has a limited application, chiefly in cylindrical joints. Expensive close-tolerance machining is necessary. The parts to be joined have to be overdimensioned with respect to the stresses induced.

1.4 Adhesive bonded joints

1.4.1 Advantages

The use of adhesives makes it possible to eliminate the weakening of the effective material cross-section since no holes are needed as in riveted

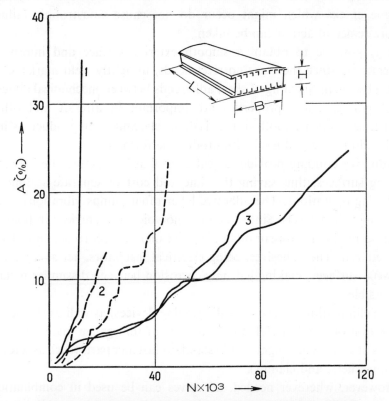

Fig. 1.1 Percentage of lost tension cross-sectional area (A) of box beams, constructed according to different designs, against the number (N) of cycles of load applied after the initiation of crack (according to results of tests performed at fluctuating stresses of $84 \pm 42 \times 10^5$ N/m², published in [1]). $L = 2,500$ mm, $B = 460$ mm, $H = 127$ mm. 1 – integrally stiffened covers machined from a plate; 2 – covers with riveted angle stiffeners; 3 – covers with bonded angle stiffeners.

and bolted joints. Bonding can be carried out at relatively low temperatures, thus avoiding the distortions of structures by heat which take place in welding, as well as structural changes and the destruction of plating or temper. The full strength of the thinnest sheets can be utilized—in high-strength materials in particular.

Due to the uniform distribution of loads over the entire area of the joint, local stress concentrations around holes and welds are eliminated, and fatigue life is increased. Fatigue cracks propagate very slowly in bonded structures (see Figure 1.1). This is of great importance, especially in aircraft, since their structures must be constructed in such a way that fatigue cracks which might occur do not cause catastrophic failure before remedial action can be taken.

It is possible to obtain a smooth exterior surface and improved appearance, since there are no rivet heads, unsightly weld marks, etc., and subsequent buckling of sheet adherends between mechanical attachment points. These properties are important in aircraft and other industries where a smooth surface is desirable, and when product styling is placed on an equal level with product performance.

Adhesive bonding provides a seal as well as a structural tie between mating surfaces, thus saving the time and cost of separate sealing or gasketing operations. The adhesive layer often damps vibrations, thus lowering sound levels. Since there are no voids and crevices in bonded joints to retain moisture or other corrosive materials, corrosion problems are reduced. The adhesives, being electrical insulators, act as a barrier between surfaces, making galvanic corrosion between dissimilar metals impossible.

The elimination of mechanical fastening devices, as well as the better utilization of the material, allows the use of thin gauge materials; this results in savings in weight and in standard fastener parts (such as screws, nuts, washers, and clamps).

However, wherever needed, adhesives can be used in combination with mechanical fasteners to seal, or distribute loads, and joints of extreme security are obtained.

The clearance between metal parts in cylindrical joints should be in the range of 0·05 to 0·15 mm, i.e. the difference in diameters is 0·1 to 0·3 mm. Thus, there is no need for time-consuming and expensive close-tolerance machining.

Not only different metals, but also different materials difficult or impossible to join by conventional methods, can be readily bonded with adhesives (e.g. glass, ceramics and china; wood, rubber and plastics; concrete and stone).

Savings due to simplified design and production techniques are achieved when adhesive bonding is employed. No skilled labour is needed in bonding operations. It is a quick and simple process lending itself to automatic methods of manufacture, which lower production time and labour costs; this is true both for short run and continuous production methods.

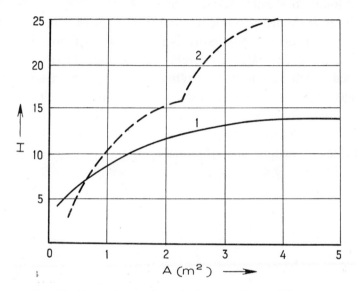

Fig. 1.2 Dependance of production man-hours (H) upon area (A): 1 – for bonded skin and stringer assembly; 2 – for identical riveted assembly.

The amount of time required for laminating by bonding is decidedly shorter than for making mechanical joints which require holding parts in a jig, or drilling holes, cutting threads, etc. Large areas and a large number of parts can be bonded in one operation. The savings in man-hours increase with the size of the component as bonding times vary little, while riveting and welding costs are directly proportional to the surface area. The curves in Figure. 1.2 show the economies achieved by bonding compared with riveting for identical assemblies. The steep rise in curve (2) reflects the additional handling time.

1.4.2 Limitations
The following are some of the shortcomings inherent to bonding: the requirement for careful preparation of the surfaces to be joined, the

need in many cases of accurate weighing of the adhesive components and their thorough mixing, the relatively long periods of time required for curing the adhesive, the need of safety precautions in handling adhesives, the lack of non-destructive methods for satisfactory evaluation of bond properties on a production basis, and the sensitivity of bonded joints to elevated service temperatures.

However, it should be borne in mind that although adhesive bonding is gaining ever more in popularity as a fastening method for metals, it still is in the early stages of development. New adhesives are being developed which have to meet requirements to withstand higher loads, for longer times, under more extreme environments. In addition to ultra-high and -low temperatures, these environments include high vacuum, ultraviolet and X-rays, attack by active fluids and fuels, high or continued rates of loading, and long time ageing.

2
Basic Principles of Adhesion

The use of adhesives is an ancient art that has become highly sophisticated; yet the problem why adhesives adhere has not been precisely elucidated. The advance of adhesive bonding will, to some degree, be dependent on the evolution of the technique from an art to a science. Despite what is known about adhesive materials, their formulations and use, the technology is still mainly supported by empirical findings.

Usually, the process of bonding comprises three stages: in the first the adhesive is applied, it sets in the second, and its properties remain constant or almost constant in the third. Due to the specificity and the diversity of the phenomena occurring at the various stages of bonding, it is very difficult to develop a theory of bonding. Moreover, an adhesive bonded joint represents in itself a complex system, in which five [1], or according to some scientists [2] even nine layers can be discerned, as shown in Figure 2.1. Each of these layers has a distinct physical size and should possess a set of unique properties.

The experimental techniques of studying the adhesion by the strength data from bonded joints may only bring confusion. On the other hand, to separate studies of adhesion from over-all joint performance is a very difficult problem in practice [3]. Firstly, this is because it is generally impossible to measure the bond strength between a plastics adhesive (as the structural ones are) and metal adherends by conventional testing methods, such as direct tensile, tensile shear and peel tests. The strength of plastics is usually lower than the strength of bonds, and in proper joints failure occurs in the adhesive layer; thus, it is not possible to determine the actual bond strength. Conversely, the interface regions between the adherends and the adhesive are physically small and experimentally inaccessible.

There is still no theory to explain in a satisfactory manner the nature of adhesion and the mechanism of the formation of a bond between the substance of the adhesive and the material of the adherends, to apply to all situations. Bonding phenomena became the object of scientific studies only some thirty years ago. In earlier investigations the main attention had been focused on the properties of the adhesive, and the materials to be bonded were considered only from a point of view of the physical condition of their surfaces. While in the past it was assumed that the bonding of two bodies was due to the mechanical anchoring of the adhesive in surface roughness and pores, it has now been definitely established that, even when bonding porous surfaces, this so-called *mechanical adhesion* is of a secondary importance compared with the *specific adhesion* due to physical and chemical phenomena.

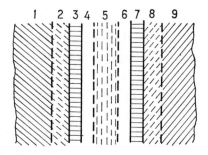

Fig. 2.1 Division of an adhesive joint into layers (in principle): 1 and 9 – bulk adherends; 2 and 8 – atomic layers of the adherends next to the interface; 3 and 7 – interfaces of atomic or molecular thickness; 4 and 6 – adhesive boundary layer affected by interfaces, having a different structure from that of the bulk adhesive; 5 – bulk adhesive layer, unaffected by the interfaces.

In later years it was stated that *chemical reactions* take place between the adhesive and the bonded surface in a number of cases. For this reason, bonding was studied from the point of view of the adhesive, as well as of the materials bonded. Moreover, a great importance was lent to surface phenomena. It was claimed that it may be possible to influence the interaction between the adhesive and the surface by adsorption layers of oriented molecules. A great importance was also attached to *wetting*. It was stated that bonding could be carried out only by adhesives that wet the bond surfaces, and that wetting is improved if surface-active additives are adsorbed on the adhesive–adherend interface.

There exist still other theories, based on different effects, which try to explain the nature of adhesion [4]. The *adsorption theory* is widely used. This theory regards adhesion as a purely surface process (similar to adsorption), and explains the formation of the bond between the adhesive and the adherends primarily as a result of the action of intermolecular forces (hydrogen bond, dipole–dipole and London-dispersion forces) [5]. The adsorption theory has been developed in the works of de Bruyne, McLaren and Staverman, among others. According to this theory, a good adhesion is possible only if both materials are either polar, or non-polar, and adhesion is impeded if one of the materials is polar, and the other non-polar. It is believed that the basic factor determining bonding is the chemical properties of the bond surfaces. The formation of a strong bond is explained as the result of the action of specific molecular forces of a physical and chemical type. More recently, Sharpe and Schonhorn have further developed the theory, explaining adhesion in terms of the relative surface tension (actually, surface free energies) of materials. They emphasize that surface tension, rather than polarity, is the property of fundamental importance in adhesion.

According to the *diffusion theory* suggested by Voyutskii [6] the adhesion consists of the diffusion of chain molecules or their segments, this resulting in the formation of a strong bond between adhesive and adherend. This theory is based on characteristic features of polymers (the chain structure of molecules, the capacity for micro-Brownian movement), and the presence of polar groups in their composition.

Deryagin and Krotova [7] are the authors of the *electrical theory* of adhesion based on the concept of the double electrical layer formed on the interface between the adhesive film and the adherend by the contact of these substances. It is assumed, that a cause for the formation of a bond is the presence of attractive forces of electrical origin, acting between the molecules, atoms and ions. Moreover, the chemical nature of the substances is of great importance. It should be stated that this theory does not contradict the adsorption theory.

Unfortunately, existing theories that explain the mechanism of bonding are still imperfect and still do not offer a possibility of relating the chemical structure of substances to their adhesion properties. And this would be of great importance, for instance, in the development of new synthetic adhesives with prescribed properties. More information on the mechanism of bonding, on the chemistry of materials, and on the design of bonded joints will lead to a well-defined science of adhesion, and enable a wider application of adhesives.

A gap exists between the power of adhesion that is theoretically possible and that practically obtainable. The factors determining the final or measured bond strength in any adhesive–adherend system, and their relation to the maximum possible or ideally obtainable bond strength, are indicated in Figure 2.2. This diagram, as published in [8], does not

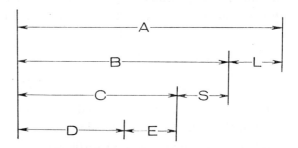

Fig. 2.2 Factors determining the strength of bonds: A – maximum possible adhesion due to interatomic and intermolecular forces of attraction; B – inherent strength of bond, determined by the ability of adhesive and adherend to attain molecular nearness; C – residual bond strength; D – measured bond strength, established at breaking the bond by externally applied stresses; L – loss caused by incomplete wetting; S – loss due to internal stress in bonds (conditions of shrinkage, thermal movement, absorption, etc.); E – loss due to defects in test or voids in adhesive layers.

depict quantitatively the relation between the various factors described.

The maximum possible adhesion A represents the ideally obtainable work of adhesion taking into account all types of physical and chemical bonding forces in the materials. Conditions to satisfy A can never be attained in practice because of the inability to produce the necessary condition of molecular nearness.

Loss L depends on the ability of the adhesive to wet the adherend. The quantitative measure of the effectiveness of wetting is the wetting angle; this depends on the surface tensions of the adhesive, the adherend and the surrounding atmosphere.

Loss S, which further reduces the bond strength, is due to internal stress development resulting from dimensional changes along the adhesive–adherend interface. This could be simply due to volume changes in the adhesive layer resulting from solvent loss, cross-linking reactions or crystallization; or a result of differential expansion at the interface caused either by thermal changes, or absorption of water or any other liquid or vapour in which the joint is immersed.

Loss E, which reduces further the maximum obtainable adhesion to the measured bond strength, is self-explanatory. The most common causes contributing to this last factor are air entrapment, specks of foreign matter, such as metal or dust, and excessive irregularities of the adherend surfaces.

3
Adhesives—Composition, Classification, Properties

3.1 Composition of adhesives

Many substances of different chemical structure may, under appropriate conditions, hold materials together by surface attachment, yet not all of them can be regarded as adhesives. For example, if water between metal or glass surfaces is frozen, a firm joint is obtained. However, ice cannot be considered as an adhesive, because it occurs only at low temperatures and is, therefore, not applicable in engineering practice.

A typical adhesive should possess a definite combination of physical and mechanical properties which make it capable of holding different bodies together without altering their structure. The basic requirements for an adhesive are:

1 At some stage of the bond formation it should be fluid.
2 While fluid, it should wet the surfaces of the adherends completely.
3 It should set to a strong solid or viscous gel.

Adhesives consist of several components [1]. One or at most two basic substances are the so-called *binders*. These are resins that are the active adhesive material, i.e. they possess the combination of adhesion and cohesion properties needed to provide the required bond strength. Some other adhesive constituents are:

1 CURING AGENTS OR HARDENERS Chemical agents used to cause the chemical process of curing the adhesive. The choice of curing agents is determined by the handling and curing techniques possible and the properties required of the cured product. They may be supplied separately, either in liquid or powder form, or may have been incorporated with the binder resin by the manufacturer; in the latter case they are activated by heat and/or pressure.

COMPOSITION, CLASSIFICATION, PROPERTIES

2 ACCELERATORS Substances which increase the curing rate of synthetic resins.

3 COLORANTS OR PIGMENTS Fine particles added to control the mixing of components, or to obtain a colour matching the adherends.

4 EXTENDERS Low-cost materials added to reduce the cost of adhesives.

5 FILLERS Solid materials, substantially inert chemically, added to modify specific properties, such as strength, viscosity, colour, or electrical properties. Typical fillers are metal powders, graphite, quartz flour, asbestos, talc, and glass fibres.

6 INHIBITORS Substances which retard the curing reaction, usually to prolong storage or working life.

7 PLASTICIZERS OR FLEXIBILIZERS Substances that do not contain reactive groups, added to lower the viscosity of an adhesive, or to increase the flexibility and toughness of a film of adhesive.

8 REACTIVE DILUENTS Low-molecular-weight compounds containing reactive groups, used to reduce the viscosity of adhesives. They are preferable to plasticizers and volatile solvents.

9 SOLVENTS Organic liquids used for the dispersion of resins, etc.

10 STABILIZERS Substances added to improve the resistance to hea or other deteriorating environments.

11 THIXOTROPIC AGENTS Substances added in order to prevent the 'sagging' of adhesives applied to vertical, inclined, or overhead surfaces. Thixotropic agents of the finely divided silica type are most common.

12 WETTING AGENTS Substances or compositions added to ensure good wetting of the adherend surfaces.

To meet the requirements for bonding applications, increasing in number and variety, a vast range of adhesives has been formulated. Cold-setting systems (curing at room temperature), as well as hot-setting adhesives (those requiring heat for curing) are available. This range includes liquid products varying in viscosity from thin fluids to stiff pastes, solid adhesives in the form of powders, pellets or rods, and adhesive films. There are no universal adhesives, and there are no simple criteria to guide the layman. Detailed advice about the effectiveness and behaviour of adhesives should be sought from the makers (see Appendix 1).

To select an adhesive for a particular application, many factors should be known and considered. These are the materials to be bonded, the bonding methods (surface preparation, adhesive application, assembly, conditions of curing), the design of the joint, and the properties required of the cured joint (for instance, the temperatures or environment which it will have to withstand). For example, if very large components are bonded, room-temperature-curing adhesives must be employed, because with heat-curing types ovens, presses or autoclaves are needed, and these are limited in size. Since there are so many factors involved in adhesive selection, a compromise may often be necessary.

There are adhesive types available to meet most bonding requirements, and new formulations are devised whenever necessary; this makes cost an important factor in adhesive selection, too. However, cost of an adhesive alone may be misleading without consideration of such factors as waste resulting from a short working life, as well as labour, inspection, and equipment requirements of any given process. In some cases a rather expensive ready-to-use adhesive may be actually cheaper in production use than a lower-cost adhesive that requires considerable labour and equipment to prepare it for use.

Attention should also be drawn to the fact that, in many cases, only a cheap cement may be needed to merely hold parts together instead of expensive structural adhesives [2]. This is valid for applications where surface preparation and joint design are less critical; where rapid setting reduces production time and tooling; where fast grab and good peel strength are more important than high shear strength.

3.2 Classification of adhesives

There are many ways of classifying adhesives, none of which, however, is satisfactory by itself [3]. It is most usual at present to classify the adhesives by their chemical composition, although many compounds represent the blending of several chemical groups, and the chemical type is often obscured by the use of trade names.

3.2.1 Classification by the materials being bonded

This method of classification is widely used, although it is of limited value to the users. Most often, tables and charts are given, listing different materials to be joined, and adhesives for this purpose are recommended, defined usually by the chemical type. Often, there are, however, many adhesives from more than one chemical group suitable for every individual case. Therefore, the relationship between the materials being joined and the chemical type of the adhesives is in a limited sense only a factor in selecting an adhesive. This relationship is of significance only for some special properties of the materials, such as porosity, strength, modulus of elasticity, coefficient of thermal expansion, and susceptibility to chemical reaction with some of the constituents of the adhesive.

3.2.2 Classification by the form of the adhesive

Adhesives are available as liquids, pastes, and in solid (dry) form as powders, rods and films. In addition to cost, the selection of the form of the adhesive is based on the type of design and the available production equipment.

Liquid and paste adhesives are those most widely used. Their viscosity depends on solvents, fillers and other constituents, as well as on temperature, pressure and other factors. Low-viscosity (free-flowing) types are the most simple to apply and are recommended for large, closely fitting surfaces. Highly filled pastes have void filling properties, and are more suitable in applications where vertical or overhead surfaces are to be bonded, or where the gap along the joint is of irregular width.

Rod and powder adhesives are usually heat-curing, and the curing agent has been incorporated by the manufacturer. They can be stored for very long periods.

Film adhesives are very convenient for bonding large areas that are flat or slightly curved; they are not suitable for components with a complicated shape. The advantage of film adhesives is that, in addition to controlling bond line thickness throughout the joint, they confine the adhesive to the immediate bonding area, and make clean bonding operations and simple application procedures possible. Since film adhesives offer many advantages, as far as design and cost saving are concerned, they are increasingly used despite their comparatively high price.

Film adhesives are made in the form of a dry film having controlled thickness and width. Supported types are reinforced with glass or synthetic fibre mats (fleeces), as well as with glass, nylon, linen or asbestos fabrics. The films may be placed on a non-adhering liner and stored in rolls. Usually they require heat and pressure for curing. The film will soften as temperature is increased, and will wet the surfaces to which it has been applied; continued heating will then induce chemical cure.

3.2.3 *Classification by bonding requirements*

Two or more of the cure requirements listed below have to be satisfied for most structural adhesives:

1. HEAT Heat-curing adhesives of all possible forms normally require temperatures up to 300°C. A higher cure temperature usually raises the bond strength, even of room-temperature-curing types.
2. PRESSURES of up to 35×10^5 N/m² are required for some adhesives, while others (epoxides, for example) do not require any.
3. TIME The time needed for a bond to attain its maximum strength depends on the temperature and pressure applied. It may vary from a few seconds to a week.
4. CURING AGENTS in liquid or powder form are supplied as one of the components of two-part adhesives. Some of the latter require additional heat and pressure for curing.

There are a number of adhesives which require no cure in the sense that setting takes place without a chemical reaction. Among these are the hot-melt adhesives which are heated to a fluid during bonding and resolidify on cooling.

Pressure-sensitive adhesives adhere to almost any surface for an almost indefinite length of time after drying. They form soft, instantaneously sticking films, and their action can be explained by a hydrodynamic mechanism.

Contact adhesives are caoutchuc-elastic substances which adhere to each other for a specified period of time after the coatings on the parts to be bonded have dried. Presumably, diffusion processes have an essential influence on their bonding effect.

A new family of structural adhesives are the one-part ready-to-use anaerobic types which harden rapidly at room temperature when deprived of air. Once applied, only that liquid which is confined inside the joint, away from air, cures.

3.2.4 Classification by chemical type

At present, this is the most common way of classifying adhesives. However, the selection of an adhesive by its chemical type, as for instance 'an epoxide' or 'a polyester', is rendered extremely difficult due to the multiplicity of chemical families of raw materials available. Another reason is the ease with which a competent chemist can manipulate these materials to obtain and make available adhesives of equivalent performance characteristics from several different chemical types. Adhesives of identical chemical compositions of binder substances are made available in a variety of physical forms. Moreover, even minor variations in the constituents used to modify the binder substance may, for instance, substantially change specific adhesion, heat resistance, or toughness, as well as curing time.

Proceeding from the type of the basic constituents, adhesives may generally be classified in two large groups: inorganic and organic.

1 INORGANIC ADHESIVES are based on inorganic products, such as sodium silicate, sodium phosphate and magnesium oxychloride. They will not be discussed, since they are not suitable for the structural bonding of metals.

2 ORGANIC ADHESIVES may be subdivided in the following groups.
 a Natural adhesives, based on animal, vegetable and modified natural products, will also not be discussed, since they are likewise unsuitable for the structural bonding of metals. Their usage is mostly limited to paper, cardboard and wood. This group includes, for example, blood albumin, shellac, proteins, starches, cellulose derivatives, rosin, asphalt, oleoresins, gums, natural and reclaimed rubber.

b Synthetic adhesives are subdivided in three main groups, and are discussed in greater detail below.

Thermoplastic adhesives are based on thermoplastic resins. They can be softened by heating and hardened by cooling without undergoing chemical changes, these phenomena being repeatable. Thermoplastic adhesives are considerably cheaper than the thermosetting types and meet many applications where low joint strengths are required. This group includes, for example, vinyl and vinyl-modified products, polyacrylics, polyamides and polyethylenes, usually used as solvent dispersions. The shortcomings of thermoplastic materials in general are typical of them, such as poor creep strength, brittleness at sub-zero temperatures and low heat resistance. For this reason, in most structural applications thermoplastic resins are compounded with thermosetting resins.

Modified thermoplastic resins known as hot-melt adhesives, though still in the early stages of development, are starting to be used for bonding metals in engineering applications when a high bond strength is needed. They have two very important advantages over conventional thermosetting adhesives: there are no problems associated with a limited working life after mixing the components, and also a bond is created rapidly without the need for subsequent operations—the solid resin is heated to a fluid, and it melts and resolidifies on cooling. Compared with the solvent-based adhesives, no trouble should be expected with solvent escapement after assembly. Other useful properties of the hot melts include a very high impact resistance and good flexibility. For example, joint strengths up to $1 \cdot 8 \times 10^7$ N/m^2 can be developed in seconds with polyamide hot-melt adhesives. Moreover, there is no difficulty in bonding plastics to themselves and to metals with hot-melt adhesives.

Elastomeric adhesives are based on synthetic rubbers, such as styrene–butadiene (SBR), nitrile, polychloroprene (neoprene), acrylonitrile–butadiene–styrene (ABS), polysulphide (thiokol) and silicone. They include not only the elastomer but also solvents, fillers and additives to impart some desirable characteristics to the end product (high flexibility, for example). These adhesives produce an instant stick when brought together; they set by the evaporation of the solvents. Very few formulations contain an elastomer as the sole adhesive ingredient, primarily because of the resulting inherently low strength. In most structural applications the elastomeric adhesives contain either thermoplastic, or thermosetting resins as an additional constituent.

Thermosetting adhesives are based on thermosetting resins which change into infusible and insoluble products as the result of a chemical reaction. The latter, generally termed as curing, is usually assisted by heat and pressure, and by a curing agent as well. Curing is a polymerization process in which polymers with large molecules are formed by linking together from smaller molecules. A distinction is made between addition polymerization yielding only the polymer, and condensation polymerization accompanied by the formation of by-products (water, for example). The thermoset materials obtained do not soften significantly on heating to temperatures below their decomposition temperature. Thermosetting resins used in adhesives include among others phenolic, urea, melamine, epoxide, unsaturated polyester, polyurethane and polyimide. Their price is generally higher than that of all other types listed above, but the bond strengths obtained are likewise correspondingly higher. Creep strength is good, but the bonds are brittle and impact resistance is low. Therefore, thermosetting resins find their widest use in serving as the basis of structural adhesives, modified with other types of resins.

3.3 Structural adhesives

The term 'structural', in the sense used here, is restricted to adhesives used for assembling load-carrying components and structures, made chiefly of metals and rigid structural plastics. This, however, is not a fixed term, since there are adhesives of all types which may be termed 'structural' in some application.

At present, structural adhesives are in principle a combination of thermosetting resins and elastomers or thermoplastics. These combinations of resins of different chemical groups are sometimes called adhesive alloys or hybrid systems. They are strong, tough, infusible, insoluble, and can be used over a wide range of temperatures.

Thermosetting–thermoplastic systems cure and behave in a manner similar to that of straight thermosetting types, but have greater flexibility, peel strength and impact resistance.

TABLE 3.1 Examples of common structural adhesive types and properties

Type of adhesive	Forms	Economics	Working life at room temperature (ready to use)	Curing conditions (RT=room temperature)	Tensile shear strength at RT ($\times 10^7$ N/m^2)	Peel strength	Impact resistance	Creep resistance	Heat resistance, °C	Solvent resistance
Phenolic-vinyl	Liq. (1 part) Liq. + powder Film	Medium	$\frac{1}{2}$ to 1 yr	150°C, 20 min Pressure	1·7 to 3·5	Good	Good	Fair	100 to 130	Fair
Phenolic – rubber	Liq. (1 part) Film	Medium	> 1 yr	175°C, 30 min Pressure	0·3 to 3·2	Good	Good	Good	175 to 230	Good
Epoxide	All forms	High	15 min to > 1 yr	RT 24 hr 300° 3 min Contact	1·5 to 4·5	Poor	Poor	Good	80 to 175	Good
Epoxide – polysulphide	Liq. (2 part)	High	15 min to 2 hr	RT, 24 hr 65°C, 20 min Contact	1·3 to 2·5	Medium	Medium	Medium	80	Good
Epoxide – polyamide	Liq. (2 part)	High	30 min to 3 hr	RT, 1 to 5 days 200°C, 3 to 5 min Contact	1·4 to 2·5	Medium	Medium	Medium	80	Medium
Epoxide – nylon	Film	High	6 hr to 1 yr	175°C, 1 hr Pressure	2·5 to 4·2	Very good	Good	—	90 to 150	Good
Epoxide – phenolic	Liq. (1 part) Film	High	1 to 15 days	165°C, 1 hr Pressure	2	Medium	Poor	Good	260	Good
Unsaturated polyester	Liq. (3 part)	Medium	$\frac{1}{2}$ to 2 hr	RT, 6 to 24 hr Contact	1 to 3	Good	Good	Med	80 to 100	Good
Polyurethane	Liq. (2 part)	High	2 to 8 hr	RT, 24 hr 175°C, 1 hr Pressure	3·5	Good	Good	Good	90	Good
Cyanoacrylate	Liq. (1 part)	Very high	$\frac{1}{2}$ to 1 yr	RT, < 10 min Contact	1 to 2	Poor	Poor	—	80	Poor
Polyacrylate	Liq. (1 part)	Very high	1 yr	RT, 1 to 24 hr 120°C, 1 min Contact	0·5 to 4	Good	Good	—	90 to 200	Good

COMPOSITION, CLASSIFICATION, PROPERTIES

When the second resin is elastomeric, the strength of the bonds is slightly lowered, as compared with rigid bonds of straight thermosetting adhesives, but resiliency and peel strength are added, as well as resistance to shock and vibration.

In practice, a wide range of properties can be obtained by combining different resins, and also by developing different formulations using the same resins. There is much research and development under way in this field.

At present, there are two main classes of adhesives that outstrip by far all others that are employed in metal-to-metal bonding: these are the modified phenolic and epoxide systems.

Types of adhesive compositions, most commonly used in structural bonding applications, are listed in Table 3.1, along with data on the form available, economic considerations, curing requirements, and mechanical properties. As stated earlier, tables of this type, giving general characteristics, cannot be of help in the selection of a specific adhesive. A reason for this is that adhesives within a class may vary widely in specific properties; moreover, inter-relationships between properties that could be of decisive importance are not shown. Therefore, such tables can only serve for a preliminary selection, or indicate the unsuitable types. A brief description of each adhesive type listed follows.

Phenolic-vinyl adhesives—The most important formulations contain a phenolic resin combined with polyvinyl formal or polyvinyl butyral. They provide good shear and peel strengths at room temperature, but their resistance to higher temperatures is limited. They are supplied as heat-curing liquids (solvent dispersed), or as unsupported and supported films. Some types are applied by coating both metal surfaces to be bonded with the liquid phenolic resin and subsequently sprinkling of the coat with polyvinyl powder, or alternatively dipping the parts into the powder. Prior to bonding, the liquid types must be air- and/or force-dried to remove volatile constituents.

These adhesives give off during the cure vapours either from solvents or, what is more important, from the products of the chemical polycondensation curing process. Owing to this, they should be cured with a pressure acting on all joint areas, which should be higher than the vapour pressure of the by-products; otherwise, at lower curing pressure, a porous bond line is formed with reduced mechanical properties. At a temperature of 150°C this pressure should be in the order of 5×10^5 N/m², if water vapour pressure only is considered. In actual production, however, the required external pressure on the joint should be higher, so that it will deform the adherends if they do not fit well, and cause the adhesive to flow and spread uniformly all over the joint. Adhesives with a low volatile and phenolic resin content may be cured at lower external pressures.

Phenolic-rubber adhesives—Phenolic resins combined with natural, reclaimed, or synthetic rubbers offer good flexibility and vibration absorption, as well as easy processing. In most formulations for structural applications, polychloroprene (neoprene) or nitrile rubber are used as the elastomeric component. Usually, these adhesives are available as liquid solutions, and as unsupported and supported films. They all require an adequate pressure during the cure, for the reasons stated in the above section on phenolic-vinyl adhesives. A wide range of curing conditions is possible. These adhesives are less sensitive to surface preparation than the phenolic-vinyls and most of the epoxies.

Phenolic-nitrile adhesives are often preferred because of their increased resistance to chemicals, oils, fuels, and solvents, as well as of their higher service temperature range; some formulations can be used for short periods at temperatures as high as 350°C. These adhesives are increasingly used in bonding a large variety of materials in the aircraft industry, and friction linings in the automobile industry.

Epoxide adhesives—Although more expensive than most other types, epoxide adhesives have found wide use for bonding metals and other non-porous materials together. This is due to their excellent strength and remarkable versatility in meeting technical requirements. These adhesives do not contain solvents and are 100 per cent reactive; therefore, there is no solvent evaporation problem when joining impervious surfaces. Since straight epoxide resins are rigid and brittle, they are usually combined with elastomeric or thermoplastic resins to improve peel strength and impact resistance. The joints resist heat to a substantial degree, and are normally unaffected by moisture or by chemicals. Heat-curing types have higher strengths than the room-temperature curing ones and maintain their strength over a wider temperature range.

Epoxide resins cure without releasing by-products in vapour or liquid form. For this reason, only contact or little pressure is required during curing. Shrinkage is negligible.

A range of different types of epoxide adhesives is available: liquids and pastes in a wide range of viscosities, solids in a wide range of melting-points, as well as unsupported and supported films, in either one- or two-part systems. They may be cured over a wide range of temperatures through proper selection of the curing agent. Curing agents commonly used in adhesive formulation include diethylene triamine (DTA), diethylaminopropylamine, polyamide resins, m-phenylenediamine (MPD), hexahydrophthalic anhydride (HPA), dicyandiamide (DICY) and BF_3. In heat-curing solid systems the curing agent is incorporated by the manufacturer beforehand.

Two-part types consist of a base binder and a separate liquid curing agent, which are mixed before use. For household needs, in particular, two-part epoxide adhesives are commonly marketed in compartmented plastic packages whose contents are combined at the time of use, or in collapsible metal tube kits from which equal-length ribbons of binder and hardener are squeezed out and mixed.

The handling characteristics of the adhesives can be modified in many ways by the formulator or the user to suit many production situations and give adequate bonds by adding various types of fillers, plasticizers, etc. [4]. A wide range of fillers may be used, such as short-fibre asbestos, aluminium powder, mica powder, aluminium oxide, talc, and graphite. Reactive diluents used to reduce viscosity include allyl glycidyl ether (AGE) and phenyl glycidyl ether (PGE). Dibutyl phthalate or another high-boiling-point esters may be used as plasticizers. Polysulphide-rubber may be used to give greatly increased flexibility to the adhesives, but at the expense of shear strength and heat resistance. Polyvinyl acetate and polyvinyl formal give increased flexibility and peel strength but this lowers heat resistance. To reduce the temperature and duration of cure, small quantities of an accelerator, such as phenol or benzyldimethylamine (BDMA), may be added.

Good strength and flexibility, as well as high water and moisture resistance have been achieved by combining epoxide resin (100 parts by weight) with modified coal tar (150 parts by weight)[5]. A fact of great importance is that the price of the adhesive is greatly reduced because of the large amount of cheap tar included. Since the tar and epoxide resin react chemically, the former should be compounded with the curing agent as one of the components of two-part adhesive systems.

Epoxide–phenolic and epoxide–silicone adhesives are common for high-temperature uses.

Polyester adhesives—These are solvent-free types based on unsaturated polyester resins compounded with monomeric substances. They are supplied as liquids in a wide range of viscosities. In addition to a curing agent, room-temperature curing types require an accelerator (three-part adhesives). Since polyester adhesives cure without giving off volatile by-products, only contact or little pressure is needed during curing. Their shrinkage, however, is considerable. The addition of adequate fillers can widely modify the properties of the adhesives.

Polyurethane adhesives—Their properties are similar to that of elastomers. However, a large variety of formulations having different hardnesses and elongations are available. The joints made with polyurethane adhesives have good peel strength and impact resistance. Resistance to chemicals, oils and fuels is good, but moisture resistance is low. They can be used for bonding metals, plastics, elastomers, foams, glass, ceramics, etc., to themselves and to each other.

One-part fast-assembly adhesives—Versatile one-part cold-setting adhesives are new developments meriting a high interest. There are at present two basic types of such adhesives: the cyanoacrylate- and the polyacrylate-based adhesives. Both types are supplied in pipette bottles and are ready for instant use. They do not require any solvents and set by 'anaerobic polymerization'. They harden in the presence of metal ions and in the absence of atmospheric oxygen without perceptible shrinkage. For this reason close tolerances on the fit of parts to be bonded are required. Contact pressure is sufficient.

Cyanoacrylate adhesives can join virtually any combination of materials and make firm, handleable bonds in seconds to minutes at room temperature. The initial bond strength is so high that there is no need for fixtures or clamps to hold the parts to be bonded together.

Polyacrylate types (ethylene-glycolacrylate, for example) need up to 24 hrs and more for hardening at room temperature; this period can be reduced, however, to minutes by heating. With these adhesives priming with special surface activators may be needed when bonding certain materials. Because of their low viscosity and the longer hardening period, these adhesives penetrate well into capillary clearances and are often applied after assembly. They are widely used in securing bearings and screws, as well as in the sealing and corrosion-proofing of joints.

The price of the fast assembly adhesives is very high at present, as compared with common structural types, thus limiting their application. For example, a typical cyanoacrylate adhesive packed in a ½-oz bottle costs as much as £3 15s 0d (£3·75). Nevertheless, these adhesives may be invaluable in detail assembly work for small bond areas.

High-temperature adhesives[6]—A distinction should be made, when assessing the thermal stability of adhesives, between strength after a short time and after prolonged exposure to high temperatures; only the latter is of interest for general engineering applications. Extremely high temperatures that should be withstood by the adhesive bond for minutes, occur only in very special cases as, for example, in space engineering when high temperatures develop on re-entry into the dense atmosphere.

For most adhesives strength falls off abruptly above 120°C. However, a number of structural adhesives retain their strength up to temperatures of about 150 to 200°C.

Although the standard bisphenol-A-based epoxide adhesives generally are not useful above 175°C, modified systems (particularly anhydride-cured systems) have reasonable strength retention at 230°C. Other types of epoxide resin combined with phenolics and novolacs can be used as adhesives at temperatures to about 250°C.

New modified silicone-based adhesives are being developed that retain their properties up to about 300°C and even above this. For higher strength adhesives, silicones are combined with epoxides and phenolics.

Polyacrylate adhesives have been recently developed that can withstand temperatures of over 200°C.

The development of fully aromatic polymers represents a major breakthrough in the field of adhesives for use at high temperatures.

They offer a great potential for making structural bonds to withstand high temperatures for relatively long periods of time. These adhesives are based on polybenzimidazole (PBI) and polyimide (PI) resins.

PI-based adhesives, while not outstanding, are good for extended use up to 350°C. They hold fast at temperatures up to 480°C for short periods of exposure. For example, the tensile shear strength of bonded stainless steel structures remains relatively unchanged at about 1·4 to $1·75 \times 10^7$ N/m² for 1,000 hr at 260°C, and then declines to about 10^7 N/m² after 2,000 hr of exposure. These adhesives are available as glass fabric-reinforced films. Curing is carried out at 175 to 290°C for 4 hr under light pressure. The price is high, but new types under development may cost less than current polyimides and half as much as PBI-adhesives.

PBI-based adhesives[7] will hold for a long time (1,000 hr) at temperatures up to 260°C. Even at 550°C they will hold for brief exposures (15 min). These adhesives generally come in film form, basically PBI coated on glass fabric. They have very high initial bond strength—on stainless steel the lap shear at 25°C is about $1·5 \times 10^7$ N/m². However, PBI adhesives have some serious drawbacks: (1) they deteriorate at temperatures higher than 300°C with prolonged exposure to air; (2) they are very high priced; (3) prolonged high-temperature cures (up to 4 hr at 200 to 400°C) and high cure pressures (15×10^5 N/m²) are required.

High-temperature ceramic adhesives that retain a useful strength at temperatures of 550°C or higher are based on a glass-type bond similar to that found in porcelain enamels and ceramic coatings. By using a frit as the basic constituent, the adhesives are applied to clean metal surfaces and fired at 550 to 1,650°C to produce the bond.

In spite of the fact that polyaromatic and ceramic adhesives are top-class products providing solutions to special bonding cases, they still are of no interest to general engineering because of their very high prices and difficult processing. Yet their development is bound to provide, at a later stage, the possibilities for their further uses in other applications, too.

Low-temperature adhesives—Below $-50°C$ most adhesives lose strength rapidly because of embrittlement. Therefore, special adhesives have been developed to give the necessary strength to bonded structures at temperatures as low as −200°C[8]. Many of the problems of bonded joints at these cryogenic temperatures are the result of stress concentrations and temperature gradients developed within the bond. However, such low temperatures are of interest in very special cases only: in space engineering, for example, where cryogenic fluids, such as liquid hydrogen and oxygen, are employed, or where low-temperature environments exist.

4
Processes for Adhesive Bonding

Successful bonding and reliable bonded joints depend on various factors. These are: preparation and pretreatment of joint surfaces (cleaning, etching, etc.), correct preparation of the adhesive (proportioning, mixing, observing the prescribed temperatures), application of the adhesive according to the supplier's instructions, correct joining of the parts to be bonded, and curing. It is also essential that certain rules concerning safety precautions for the working personnel are strictly observed.

4.1 Surface preparation

Since adhesion is a surface phenomenon, the methods used to prepare the surfaces of the adherends are vital to the success of a bond. However, because the nature of bonding (of adhesive forces) has not yet been elucidated, it is at present not possible to determine on the basis of theoretical knowledge the most suitable surface preparation method for each specific case.

The increase in bond strength with rough surfaces is explained by the increase in the bond surface. In practice, three types of surface should be distinguished:

1. The geometric surface, defined by the geometric dimensions of the bond area.
2. The actual surface or microsurface, defined by surface roughness.
3. The effective surface, defined by the part of the microsurface actually wetted by the adhesive.

Whereas it is possible to determine analytically the actual surface on the basis of microroughness measurements, or experimentally by gas absorption measurements, no method is known at the present for determining the effective surface. In practice, however, all data for bonded joints are related to the geometric surface.

Any metal, glass or other surface, exposed to normal atmospheric conditions, is covered by a water film, along with which different gases and vapours are bound due to absorption. Thus, in a short time every surface is covered by a layer reducing the adhesion. Furthermore, greasy layers are formed and contaminants are deposited on the surfaces of materials during transport and storage.

In order to obtain maximum bond strengths, a clean surface is usually required in order to provide proper wetting. A simple test for this is to spread water over the surface. A clean surface will hold a continuous, break-free water film; a contaminated surface will hold the water in the form of droplets or puddles. Since the surface tension of water is higher than that of most plastics, if the water drop spreads over the metal surface, it can readily be assumed that the adhesive will wet the surface, too. A more reliable variation of this method is the measurement of the surface area of water drops of exactly fixed volume when placed on the prepared metal surface.

Some general guide rules for proper surface preparation will be given below, followed by typical techniques for handling particular materials.

4.1.1 Precleaning
The first step is to remove loose deposits such as dirt, scale, flaking paint, and any foreign matter that impedes the wetting of the base material.

4.1.2 Mechanical surface pretreatment
Mechanical pretreatment methods include blasting, abrading, machining and scouring with sandpaper. Blasting is the preferred pretreatment method for metals, and only when metal members are of such shape or thickness as to make blasting inpracticable (for example, because of warpage danger), chemical etching is recommended. While usually all metals may be blasted for about 30 sec with sharp and clean steel or iron grit (size 0·25 to 0·4 mm), for aluminium only sand or aluminium oxide should be used. Areas that are to receive no adhesive should be masked out. Steel wool or wire brushes are used for abrading. The best results

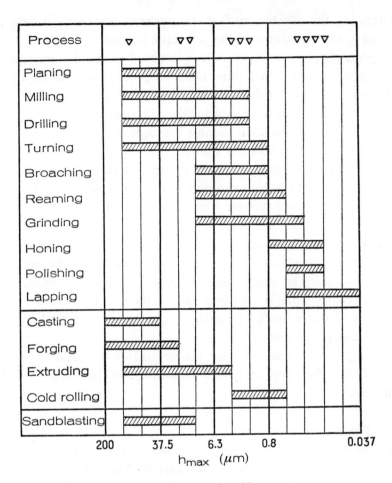

Fig. 4.1 Surface roughness produced by common production methods (h – roughness height).

are obtained by cleaning down to bare metal. All dust should be removed before bonding.

Figure 4.1 illustrates the range of surface roughness normal to common production methods. The effect of surface roughness on the strength of bonded joints is shown in Figure 4.2. The strength is increased due to an increase in the effective bond area. However, roughness height should not exceed a certain limit. A surface that is too rough can weaken rather than strengthen the bond because contact between surface peaks may occur resulting in breaking of the adhesive layer. Furthermore, the

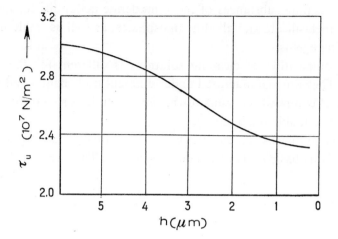

Fig. 4.2 Effect of surface roughness height (h) on the tensile shear strength (τ_u) of aluminium specimens bonded with a heat-curing epoxide adhesive.

coarse surface is more difficult to wet completely, and requires thick layers of expensive adhesive.

4.1.3 Surface degreasing

It is essential that the materials to be bonded are free from grease. It is often claimed that even a thumb-print on an otherwise clean surface will impair adhesion.

Organic chemicals commonly employed for degreasing bonding areas include trichloroethylene, perchloroethylene, methylene chloride, acetone and petrol. It is desirable that degreasing is not simply carried out by wiping, since dissolved grease is collected in the solvent, as well as on the cloth or brush used; there is the danger that a grease film will be left on the metal surface after evaporation of the solvent. Immersion in solvent baths is not recommended, since the amount of dissolved or suspended soil rapidly builds up, and the parts are recontaminated. Best results are obtained by degreasing in a vapour bath. For example, immersion for 30 sec in a trichloroethylene vapour bath is usually sufficient for complete degreasing. This method is especially economical for either cleaning large surfaces, or large quantities of small parts.

Inorganic alkaline solutions are also used in industry for degreasing. In addition to sodium hydroxide and sodium carbonate, they frequently contain alkaline substances of more moderate action as, for example, sodium metasilicate and alkaline phosphates, as well as small amounts of wetting agents.

The composition of cleaning solutions is different for the various metals. They are manufactured under different trade names and usually their exact composition is unknown. Some solutions also form a strong oxide film suitable for bonding.

Special emphasis should be laid on castings and forgings used frequently as basic parts of bonded assemblies. The cleaning procedure

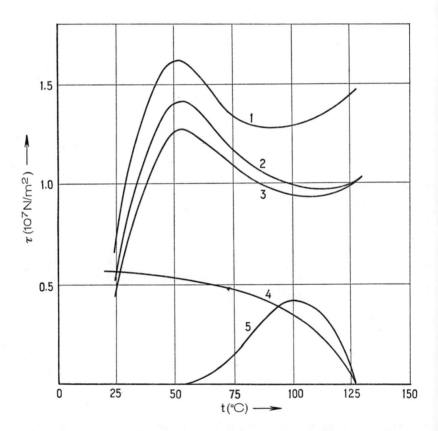

Fig. 4.3 Influence of the curing temperature (t) on the strength (τ) of adhesive bonds to contaminated aluminium surfaces. Adhesive: two-part room-temperature-curing epoxide. Contaminants: (1) no contaminant (for control); (2) machine oil; (3) paraffin oil; (4) stearine acid; (5) lubricant grease

Note. Before the application of the contaminants under controlled conditions, the metal surfaces were sandpapered, and then wiped with acetone.

is necessarily different from that used for wrought sheet materials. If the castings are cleaned by immersion without application of ultrasonics, it would be extremely difficult, and probably impossible to remove all the cleaning solution retained in small voids and openings by rinsing the casting after cleaning. Grit- or sand-blasting of the surfaces to be bonded has been found to be more successful.

It should be noted, however, that under certain conditions it is possible to bond parts with contaminated surfaces and achieve nevertheless satisfactory results. Some results of experiments to prove this[1] are given in Figure 4.3. It is seen that the loss in bond strength for some contaminants is surprisingly low, especially when the adhesive is heat-cured. Thus, it is possible to give up in some applications the sometimes inconvenient and expensive surface pretreatment procedures when maximum strengths are not particularly required. For example, cold-rolled steel or cast iron parts may be bonded, even if they still have the protective oily film used during fabrication. This may be explained by the plasticizing effect of the hydrocarbon oil on the adhesive which increases the peel strength more than it impairs the tensile strength. And, as it will be shown in Chapter 5, tensile shear strength results from a complex combination of both tensile and peel action.

4.1.4 Chemical pretreatment of metal surfaces

Chemical methods consist of the application of organic or inorganic reagents which alter the surface composition and/or increase the surface area. In most cases, the use of these methods results in significant strength increases. However, their use is expensive and requires close control.

Many metallic surfaces tend to form oxides, especially at high temperatures. Numerous metal oxides, however, have very weak cohesive strength in themselves, and some metals tend to form excessively thick oxide layers. Therefore, when an adhesive bond is made to these oxides, failure will occur in the oxides at relatively low load levels.

Oxide films can be removed by mechanical means such as machining or abrasion, but instantly a new oxide film starts to form. There are acid and alkaline solutions which are effective for the removal of oxides. However, after having been rinsed and dried, the parts have a new oxide film; but the new film may offer new properties which are often more desirable for a specific application.

The usual etching solutions used are acids or bases that attack the oxides more than the base metal. They permit a selective oxide formation essential to the formation of good adhesive bonds. Thicker oxide films are removed by pickling. Milder solutions that contain dilute concentrations of acid plus chromate additives, may be used for further etching. These leave thin films which retard subsequent oxidation and promote wetting by adhesives. Thorough washing and agitation (by mechanical means, ultrasonics, or oil-free compressed air) in any of the baths is essential. Subsequent rinsing and drying is just as important since weak bonds may result from solution entrapment. Pure and soft water is required for rinsing, as well as for all treatment solutions.

Metals that resist oxidation are readily bonded without previous etching. This is valid for the noble metals and brass, for example.

Some typical chemical pretreatment procedures for metals and the corresponding etching compositions are given below (all percentages are by weight):

1 LOW-CARBON STEELS

Sulphuric acid treatment—etch for 10 min at 80°C in the following solution: 17% conc. H_2SO_4, 83% water. Neutralize with soda solution and rinse with methanol.

Sulphuric – oxalic acid treatment—etch for 30 min at 60°C in the following solution: 10% conc. H_2SO_4, 10% $(COOH)_2$, 80% water. Rinse with cold water and remove the oxide layer by steam jet.

2 STAINLESS STEELS

Sulphuric – oxalic acid treatment—as for low-carbon steels.

Hydrochloric acid treatment—etch for 15 min at room temperature in the following solution: 30% conc. HCl, 70% water. Wash thoroughly in tap water and rinse with distilled water.

3 ALUMINIUM AND ITS ALLOYS

Sulphuric acid – sodium dichromate treatment (Pickling process according to British Aircraft Process Specification DTD 915B)—pickle for 30 min at 60 to 65°C in the following solution: 27·5% conc. H_2SO_4, 7·5% $Na_2Cr_2O_7.2H_2O$, 65% water. Wash thoroughly in tap water and rinse with distilled water.

Alkaline treatment—each in inhibited caustic soda lye according to supplier's instructions.

4 MAGNESIUM AND ITS ALLOYS

Nitric acid – potassium dichromate treatment—etch for 1 min at room temperature in the following solution: 20% HNO_3, 15% $K_2Cr_2O_2.2H_2O$, 65% water. Wash thoroughly in tap water and rinse with distilled water.

5 COPPER AND ITS ALLOYS

Sulphuric acid – sodium dichromate treatment—pickle for 5 min at room temperature in the following solution: 27·5% conc. H_2SO_4, 7·5% $Na_2Cr_2O_2.2H_2O$, 65% water. Wash thoroughly in tap water and rinse with distilled water.

4.1.5 Anodizing

Good strengths have been attained when bonding structures of anodized details. However, the pretreatment methods using acid etchants are superior to anodizing, and the latter is seldom recommended as a surface preparation for bonding, subsequent to pickling. Anodizing is prescribed, for example, in the British Aircraft Process Specification DTD 910C.

4.1.6 Surface preparation of miscellaneous materials

Plastics—Moulded plastics articles and plastics laminates should be treated by sandpapering or by a light sand-blast. The purpose is to remove the smooth glossy surface since it may contain either parting or mould-release agents. An acetone or methanol wipe may follow. The bonding of some thermoplastics, (such as polyethylene, polypropylene and fuorcarbon polymers) presents difficulties; they require etching to modify the surface, or flame treatment using an oxygen-rich flame.

Rubber—Best results are obtained by cyclizing or etching the surface. This is done by immersing the surface to be bonded in concentrated sulphuric acid—for 5 to 10 min in the case of natural rubber and 10 to 15 min in the case of synthetic rubber. After thoroughly rinsing with water and drying, the brittle surface of the rubber should be broken by flexing, so that a finely cracked surface is produced. To neutralize any remaining acid, a wash with diluted caustic soda solution (0·2%) is sometimes recommended. Some soft or foamed rubbers containing small amounts of non-polar filler need only to be cleaned with benzene.

Glass should be pickled as for aluminium or abraded.

Wood should only be smoothed with glasspaper. It must be kept free of oils and greases. Best results will be obtained from both a dimensional and strength standpoint if the moisture content is about 10 per cent. No attempt should be made to clean further after sanding.

Ceramics, *leather*, and other materials should be abraded and degreased.

4.2 Preparing and applying the adhesive

Adhesives are supplied in different forms. They may consist of several components, such as binder, curing agent, solvents and fillers. The adhesives in solid form (rods, powders, films) contain already all constituents, while in liquid adhesives it is necessary first to proportion and mix the components and additives according to the supplier's prescriptions.

4.2.1 Mixing the adhesive components

The method of mixing depends on the adhesive type. While some adhesives require the observance of accurate mixing proportions, others allow considerable deviations. Paste adhesives are proportioned usually by weight, whilst liquid adhesives may be proportioned by volume. For most epoxide adhesives, for example, it is very important to mix the constituents in exact proportions. The hardener-addition tolerance limits are ± 5 per cent, and an excess of any component will reduce the bond properties. While an insufficient amount of hardener will fail to help polymerize the mixture completely, too much hardener will produce a brittle bond; unreacted excess of amine or anhydride hardeners may also cause corrosion of adherend surfaces.

For very small volumes of a thin resin system, resin and hardener can be proportioned from a simple burette. Another simple method of proportioning is just by weighing the components out of their containers and into the mixing cup. With the easy-to-handle 'one-to-one mix' (by volume) adhesives there is no need for scales or special equipment and the time-consuming weighing of separate components is eliminated.

Heat-curing adhesives may usually be stored for long periods of time

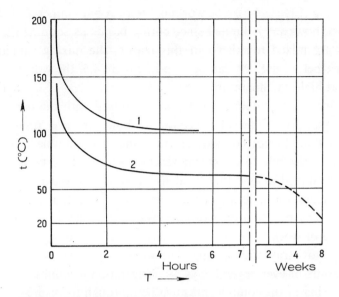

Fig. 4.4 Relationship between working time (*T*) and temperature (*t*) for a heat-curing epoxide adhesive after mixing the components: 1 – for small quantities; 2 – for large quantities ($\approx$ 0·5 kg).

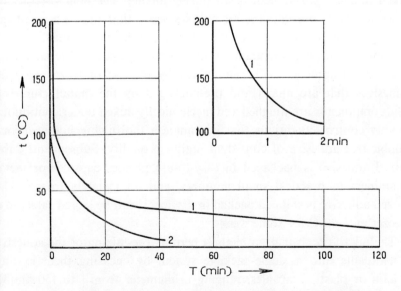

Fig. 4.5 Relationship between working time (*T*) and temperature (*t*) for a room-temperature-curing epoxide adhesive after mixing the components: 1 – for small quantities (< 20 g); 2 – for large quantities (> 0·1 kg).

after mixing (Figure 4.4). The working life of room-temperature-curing adhesives, however, is limited since curing begins as soon as the components are mixed together; for this reason, the mixing should take place just before use. As can be seen in Figure 4.5, the working life is shorter at higher temperatures.

Adhesives that contain solvents may be mixed with the hardener in greater quantities. Special precautions, however, are needed with adhesives that do not contain solvents, since a noticeable exothermic temperature rise due to the curing reaction will result when more than 200 gs are mixed at one time, and the mixture allowed to stand in a compact mass for a longer period of time (more than 15 min, for example). The build-up of excessive exothermal heat with consequent shortening of working life can be largely offset by pouring the adhesive into a shallow tray after mixing. The working life can be extended to several hours or even several days by refrigeration techniques.

The mixing of the components must be thorough to ensure a uniform distribution of the hardener in the resin. To control the mixing the resin and the hardener may be pigmented in different colours; mixing is continued until a uniform colour is obtained. In addition, care should be taken to avoid entrapment of air during mixing. The high viscosity of most adhesives prevents the air bubbles separating and as a result inhomogenous adhesive layers with a low strength are obtained.

Improper in-plant mixing of resin and curing agent, material waste, and air entrapment can be eliminated by the use of frozen epoxide adhesives that are mixed and pre-packaged by the manufacturer[2]. The components are weighed and mechanically mixed under continuous quality control conditions that guarantee a thoroughly homogeneous bubble-free adhesive of consistent, uniform quality. Subsequently, the mixed adhesive is packaged in easy-to-use plastics capsules or metal containers in practical small amounts, and immediately frozen. The frozen adhesive is shipped packed in dry ice. It can be stored safely in a freezer at $-40°C$ for some weeks.

The micro-encapsulation method permits pre-mixing of the adhesive constituents into a single-package system by encasing them in tiny gelatin or plastics capsules, ranging in diameter from 1 to 150 μm[3]. The pre-mixed system can be stored indefinitely because the active components are separated by the walls of the microcapsules. When the adhesive is to be used, it is applied as a dry powder or as a fluid dispersion. By application of pressure or heat, the walls of the capsules are broken or melted and the active components are combined at the place.

For specific purposes, fillers such as metal, glass or porcelain powders, graphite, asbestos, aluminium oxide, and many others, may be added to the adhesive. They must be entirely dry and not contaminated by oils, waxes and fatty acids. It is recommended that the resin and the fillers be mixed first, and then the hardener added just before use. When working with old resins, the mixing of the fillers should be carried out at high temperatures.

In choosing the containers for mixing, cost and the possibility of repeated use must be taken into account. It should be borne in mind that the good adhesion of the adhesives shows up well on all utensils which make contact with them during the processing. To reduce sticking, all vessels and containers, especially metal ones, should be coated with silicone laquers or be chrome-plated. It is important to ensure that the equipment is thoroughly cleaned after use with solvents (with acetone or cellulose thinners for example) before the resin mixture remains have cured.

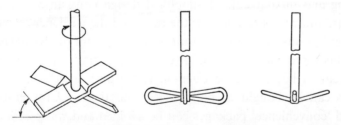

Fig. 4.6 Different types of agitators.

For very small volume users, hand mixing is usually best. For epoxide resins the use of pliable polyethylene or rubber cups is recommended since these allow cured residues to be removed by flexing. Disposable metal, glass or cardboard containers are equally suitable for mixing. Metal spatulas, paint paddles, wooden tongue depressors, etc., can be used for mixing, provided the mixture in the corners of the containers is not missed.

Many small and medium volume (0·5 to 50 litres) users find an electric hand drill ideal for mixing the components. The proper size agitator need merely to be inserted into the chuck of the drill, which should run at speeds up to 100 rev/min. Agitators can be elaborate propellers or turbines, or simply ¼-inch steel rods bent into a hoop or square shape (Figure 4.6). Egg beaters with an extended shaft have also been found very efficient.

For production installations, continuous and automatic metering and mixing equipment is available [4]. Figure 4.7 shows as an example a

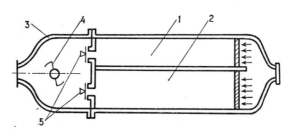

Fig. 4.7 Metering and mixing unit for two-part adhesives: 1 – hardener compartment; 2 – resin compartment; 3 – mixing chamber; 4 – agitator; 5 – regulating valves.

metering and mixing unit of simplified design consisting of a mixing chamber, and compartments for the resin and the hardener, placed in a common housing; the joint bottom of the resin and the hardener compartments represents the piston of the injection unit [5]. By using such equipment in large-scale production, resins and other adhesive components can be bought at bulk prices, and the need for costly pre-weighed 'convenience' packages can be avoided and waste reduced. In this case, however, the problem of the optimum mixture arises. To this end it is very expedient to use a simple method worked out by German workers [6] that requires no statistical knowledge or much time. This method will be described briefly.

Usually, maximum and minimum values are given for the quantity of each constituent of a mixture. In practice, however, unwillingly, mean quantities are nearly always taken. But if the price of the constituents is different, an optimization would be worthwhile. For example, eight substances should be mixed within the limits given in Table 4.1, where they are arranged in accordance with increase in price. Table 4.2 is then drawn up. In the first line the prescribed minimum quantities are

PROCESSES FOR ADHESIVE BONDING 41

TABLE 4.1 Permissible mixing ratios and prices

Constituent	Minimum, %	Maximum, %	Price, shilling/kg
A	0	17	6
B	2	10	8
C	1	30	10
D	0	13	16
E	4	12	17
F	3	40	20
G	1	30	22
H	2	10	26

entered and are summed up in the last column. In this particular case the sum comes to 13 per cent of the mixture, and 87 per cent remains to be used. To obtain the cheapest mixture, as much as possible of the cheapest constituents should be used, certainly within the range of the maximum percentages permitted. However, one part of the maximum quantity is already contained in the first line. Due to this, the differences between the maximum and minimum values are entered in the second line. At the same time the entries are summed up, until the sum of the percentages in the second line reaches, approximately, the value of 87 per cent. At E, 75 per cent is attained. There is 40 per cent of the substance F available, only 3 per cent being already used up; however, only 12 per cent is needed to attain 100 per cent, and this value is entered in the third line. The fourth line is obtained by summing up, thus serving as a control to see if 100 per cent is reached.

TABLE 4.2 Determining an optimum mixture

Line	Constituent								Sum
	A	B	C	D	E	F	G	H	
1	0	2	1	0	4	3	1	2	13
2	17	8	29	13	8	—	—	—	75
3	—	—	—	—	—	12	—	—	12
4	17	10	30	13	12	15	1	2	100

Example—A polyester resin based adhesive is to be mixed, having the desired mechanical properties and a curing time corresponding to the production process. The mixture should be prepared so as to obtain a product lowest in cost, possessing adequate properties (guaranteed within the minimum limits). The minimum and maximum quantities and the prices are given in Table 4.3.

D

TABLE 4.3 Permissible mixing ratios and prices of adhesive constituents

Constituent	Minimum, %	Maximum, %	Price, shillings/kg
Asbestos flour	0	40	1
Thixotropic agent	0	30	3
Polyester resin	56	99	5
Accelerator	0·3	4	18
Curing agent	0·6	4	25

The minimum quantities are entered in the first line of Table 4.4. Their sum is 56·9 per cent. The differences between the first line and the permitted maximum quantities are then entered in the second line until coming close to the still unused part of $100-56·9=43·1$ per cent. Since asbestos flour participates with 40 per cent, this value is entered. The third line has only 3·1 per cent left, that is covered by the thixotropic agent. All values are summed up for checking. Then, the only thing left is to see whether the adhesive obtained meets to all mechanical and processing requirements.

TABLE 4.4 Determination of the optimum mixture for an adhesive

Line	Asbestos flour	Thixotropic agent	Polyester resin	Accelerator	Curing agent	Sum
1	0	0	56	0·3	0·6	56·9
2	40	—	—	—	—	40·0
3	—	3·1	—	—	—	3·1
Cheapest adhesive	40	3·1	56	0·3	0·6	100·0

A formal mathematical solution using a linear optimization technique would require the working out and further calculation of a numerical scheme with 17 by 7 fields, for which the use of a computer would be justified.

4.2.2 Application of the adhesive

Proper adhesive application is important to obtain maximum joint properties. Improper adhesive application techniques can result in partial or complete failure of an assembly. Care should be taken to avoid contaminating the adhesive and the cleaned metal surfaces by any substance which will hinder the wetting action of the film. It is advisable to bond the freshly cleaned surfaces immediately or within 8 hr of surface preparation. In some cases primers are applied to protect the cleaned parts until bonding can be completed.

When bonding porous or semi-porous materials, it is sometimes desirable to obtain a deeper penetration of the adhesive to obtain a better bond; this may be accomplished by priming these surfaces before the application of the adhesive.

Parts or surfaces that are not to be bonded should be covered before bonding with greasy paper, polyamide foils, or coated with wax, paraffin, or other parting agents.

Room-temperature-curing adhesives—These adhesives are usually supplied in liquid or paste form. Over small areas a thin coat of the mixed compound should be applied by hand using a brush, a roller, a wood or metal spreader, or similar conventional equipment. For large surfaces spraying is the most practical application method. There are several spray-gun types of equipment, often called 'catalyst' guns, which are used in the reinforced plastics industry. These guns are specifically designed to mix low-viscosity resins and hardeners in an atomized state, and deposit the mixed system as a liquid coating. Since air is the carrier, high-viscosity systems tend to retain air bubbles after spraying; owing to this they should be thinned. Proportioning units can be used to feed the spray guns. Ordinary spray guns can also be used for previously mixed systems. Their use, however, is limited by the short working life after mixing the adhesive and the difficulty of cleaning of the guns. Usually, the working pressure of the spray guns is 0.7 to 3.5×10^5 N/m², depending on the viscosity of the adhesive. They are held at a distance of about 25 cm from the bonding surface and moved at an even speed. It is advisable to apply two thin layers, the second movement of the gun being perpendicular to the first movement.

If the adhesives contain solvents (or thinners), a suitable period of time should be allowed to elapse for their evaporation. This does not apply strictly when one of the materials is porous since the solvents are then absorbed. If the solvent remains in the bond line this will result in strength loss. Volatile components produce bubbles and pores that may be the centres of internal stress concentrations, particularly if the pressure of volatile component vapours is comparatively high. A correctly air-dried surface should be tack-free. Solvent evaporation may be accelerated by force-drying.

The adhesive layer should be applied uniformly so that no air bubbles are trapped. Voids or air pockets in the bond line introduce weaknesses that endanger the integrity of the bond. Figure 4.8 shows an example of an incorrectly applied adhesive layer, causing air entrapment. Void problems may be solved by employing vacuum.

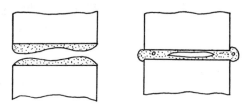

Fig. 4.8 Air entrapment caused by the unfavourable form of the adhesive layer.

If the surfaces of the adherends are irregular, a layer of glass cloth may serve as a carrier building up the bond line and securing the required thickness.

For quick fastening with room-temperature-curing epoxide adhesives a combination bonding technique has been developed [7]. A drop of a cyanoacrylate-based adhesive is surrounded with a layer of epoxide adhesive. A box formation, as illustrated in Figure 4.9, should be used for best results. Clamps and jigs, generally required with epoxide adhesives, may not be needed since the cyanoacrylate adhesive sets within minutes of contact pressure, securing the bond until the full strength of the epoxide develops.

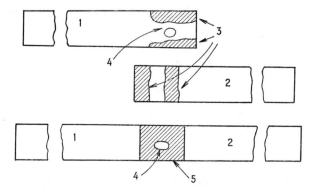

Fig. 4.9 Double adhesive system: 1 – first specimen strip; 2 – second specimen strip; 3 – room-temperature curing epoxide adhesive; 4 – one drop of cyanoacrylate adhesive; 5 – thin film of cured epoxide adhesive.

For assemblies that require a continuous seam or bead, hand- or air-operated applicator flow guns are very useful. They are fairly inexpensive and use polyethylene cartridges with a variety of different shapes of nozzles for different jobs.

Much cheaper all-plastics hand-operated syringe systems handle volumes up to 10 cm³. They are ideal where only a very small volume (one drop or two) is needed per assembly, and can be thrown away after use.

Heat-curing adhesives—Liquid heat-curing adhesives are generally applied as described in the foregoing section. Another method which makes possible the achievement of astounding speeds and efficiencies is the so-called curtain coating, where the materials are fed through a cascade of adhesive. This method is very suitable for flat parts. Adhesives used for curtain coating must be adequately formulated to maintain a critical balance between viscosity, flow properties, and surface tension so as to keep up continuity of coating, yet not thicken and build up on the weir. The construction in principle of a curtain-coating unit for automatic application of a uniform adhesive layer to sheet metal parts is

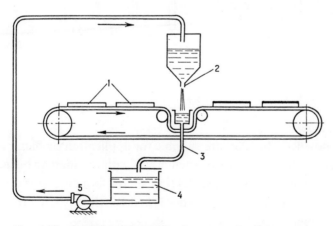

Fig. 4.10 Curtain-coating unit; 1 – parts; 2 – flow nozzle; 3 – surplus adhesive; 4 – adhesive supply; 5 – pump.

shown in Figure 4.10. The parts are conveyed through the coater at a speed of 5 m/min for example. Surplus adhesive falls into an overflow chute and is re-circulated through the coater. Evaporated solvent is replaced automatically to maintain the proper adhesive viscosity.

Solvent-containing adhesives should be dried after application until free of solvents, to a tacky state. After drying, the adhesive-coated components may, if necessary, be left for a considerable time (up to several weeks) before bonding; they should, however, be protected from contamination (by wrapping in paper, for example).

Adhesives in rod form are applied to the bonding areas after the parts to be bonded are heated to the required temperature. The necessary amount of adhesive is melted from the tip of the rod onto the bonding surfaces.

There are several methods of applying powder adhesives on the bonding surfaces: (1) they may be dusted and then sintered by heating; (2) the parts can be dipped into a melt of the powder; (3) the parts are heated to the required temperature and are dipped into the powder, or the powder is dusted onto the heated surfaces. To obtain an even layer it is desirable to sift the powder so that a uniform grain size is obtained.

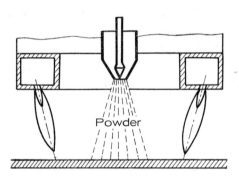

Fig. 4.11 Annular burner of a flame-spraying gun.

Some powder adhesives are suitable for application by flame spraying from a gun. The front part of such a gun with an annular burner [8] is shown in Figure 4.11. A pre-heating of the bonding surfaces is not needed in this case since the powder is melted while passing through the annular flame.

Adhesives, such as epoxide, in the form of meltable pre-shaped pellets are suitable for mass-production assembly of small parts, since they fit into the assembly like any other part [9]. They are placed exactly where needed and in the correct amount, so that no subsequent cleaning is necessary. Another advantage is that different types of adhesive can be easily included in the same assembly; for example, at one point a high-strength adhesive may be required, and at another point the adhesive may have to conduct electricity.

The only preparation necessary for adhesives in film form is the cutting to the shapes of the detail parts which are to be bonded with any of the usual cutting tools. Care must be taken that the film does not become contaminated by oils and dirt. For this reason protective layers should not be stripped off before cutting. The film adhesive is stuck to the adherends by heating, or wetting with a solvent. In some cases it may be desirable to dissolve the film adhesive and apply it by spraying.

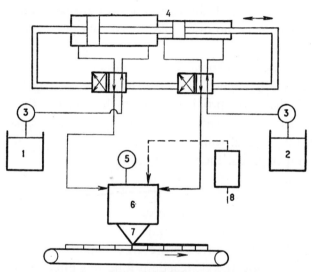

Fig. 4.12 Automated system for applying a two-part adhesive: 1 – base resin: 2 – curing agent; 3 – pumps; 4 – measuring cylinders; 5 – air motor; 6 – mixer; 7 – flow nozzle; 8 – solvent flush.

Many adhesive-application operations can be fully or partially automated, and semi-automatic and automatic proportioning, mixing and dispensing units and adhesive-application equipment can be relatively easy integrated into a continuous production system. An appreciation of the progress being made in automated adhesive assembly can be obtained by a study of the examples described in [10]. Figure 4.12 shows, for instance, a typical automated system for preparing and applying a two-part epoxide adhesive to production line parts. The base resin and the curing agent are each supplied by pumps from separate containers to separate but coupled four-way valves and coupled double-acting measuring cylinders. As the piston of the cylinders move back and forth, resin and curing agent are taken into one chamber of each cylinder, and are exactly metered out from the other cylinder chamber to the outlet port of the valve and then to an air-operated mixer. After mixing, the adhesive is applied to the parts through a pressure flow nozzle. The operation can be made continuous if the parts are placed end to end on the conveyer line, or the adhesive application could be automatically turned on and off if the parts are spaced.

Hot-melt adhesives—Solid hot-melt adhesives are heated to 150–200°C to a fluid which melts and resolidifies on cooling. An adequate pressure is usually required.

Usually the metal components to be bonded are heated before the molten adhesive is applied. This can be a disadvantage with certain assemblies if the whole product has to be heated. But the use of hot-melt adhesives can be expected to increase with the development of heating techniques that will apply the heat to the bond area only, and ovens are not required. Induction heating is proving very satisfactory here.

4.2.3 Precautions in handling adhesives

Cured adhesives are generally harmless from a physiological point of view. Most uncured adhesives, however, contain chemical components that may cause severe allergic reactions following direct contact, inhalation or ingestion, if certain precautions normally taken when handling chemicals are not observed. This is true for instance of the formaldehyde used in phenolic adhesives, the isocyanates used in polyurethane adhesives, and most curing agents. Uncured materials must not be allowed to come into contact with foodstuffs or food utensils. Measures should also be taken to prevent the uncured materials from coming in contact with the skin, since they may cause dermatitis in people with a sensitive skin.

To ensure maximum safety, the following rules should be observed when handling adhesives:

1 The working personnel should first of all have a thorough knowledge of the dangers typical of the job.

2 The areas in which adhesives are handled should be kept clean. The working surfaces should be covered with paper or plastics sheeting which can regularly be changed. Disposable paper towels—not cloth towels—should be used to clean table tops, equipment, hands, etc.
3 The working areas must be adequately ventilated, if possible locally.
4 For the individual protection of the skin the use of barrier creams or rubber gloves is advised. Touching sensitive parts of the body with contaminated hands should be avoided; however, if accidental skin contact occurs, immediate washing with soap or cleanser and water is imperative. The skin should be thoroughly cleansed at the end of each working period either by washing with soap and hot water, or by using a resin-removing cream. The use of powerful solvents, such as acetone and alcohol, to clean hands should be avoided, since they cause the skin to dry and chafe; eventually these dry areas will crack open. Laboratory coats or overalls should be worn; these should be regularly laundered.
5 Some chemicals used in adhesives, especially solvents, are highly inflammable. This calls, when working with them, for an observance of fire-safety precautions.

4.3 Joining the parts to be bonded and curing the adhesive

For all adhesives there exists a time–temperature relationship, upon which the curing cycle is based. This is shown in Figures 4.13 and 4.14 for two epoxide-based adhesives; in the second case it is illustrated how curing conditions influence bond strength. Higher temperatures greatly speed up the rate of cure and also give somewhat better properties, but it should be borne in mind that maximum strength is not obtained by quick curing.

In most cases the minimum limits of the specified heat-cure temperature and time are the most important. An extension of the cure time or a higher curing temperature (within reasonable limits) is much less critical, and usually results in brittle adhesive layers and reduces the impact resistance of the joints. Below the minimum limits the curing reaction cannot be fully completed and the optimum mechanical properties will

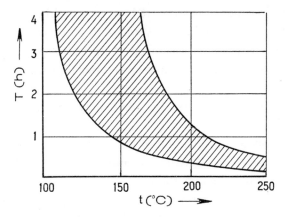

Fig. 4.13 Dependence of curing time (T) on curing temperature (t) for a heat-curing epoxide adhesive. The zone of high quality bond is shaded.

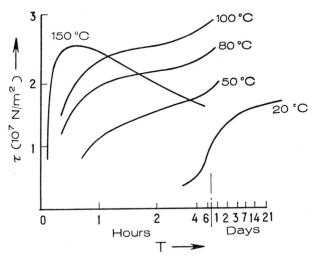

Fig. 4.14 Relationship between bond strength (τ), curing temperature and curing time (T) for a room-temperature-curing epoxide adhesive.

not be achieved at all. Overcure, however, will be detrimental to the mechanical properties only if the specified temperature and time ranges are passed considerably. When curing at room temperature, maximum strength may be attained only after several weeks.

Cure time depends upon the cure temperature, methods of heat application, production limitations and the bond properties required.

Since no two operations are exactly alike, it is advisable to conduct always some simple experiments varying both temperature and cure time to determine the optimum conditions for each particular operation.

Heat-curing adhesives should not be used for bonding assemblies if the relative coefficient of thermal expansion of the components and the adhesive differ considerably. The stresses set up in the joint during curing may increase to such an extent as to cause failure of the bond.

Heat curing can be carried out by the use of conventional infra-red, electrical resistance or gas heaters, heating blankets or tapes, autoclaves, laminating platen presses, laboratory-type or conveyor-type ovens. It is of primary importance that the temperature should be uniform over the whole bond area, and instruments for close temperature control during the cure are needed.

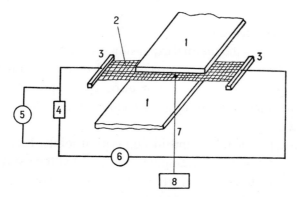

Fig. 4.15 Internal electrical curing of adhesives: 1 – adherends; 2 – resistance heater incorporated into the adhesive; 3 – busbars; 4 – power source; 5 – voltmeter; 6 – ammeter; 7 – thermocouple lead; 8 – temperature recorder.

A recent development in electric curing of adhesives [11] is illustrated in Figure 4.15. High-resistance conductive elements are placed in the bond line; heat for curing is generated when electric current passes through these elements. This results in increased production rates, since cure temperature is reached quickly. Another advantage is that large assemblies can be bonded in a room-temperature environment without the need for large ovens or autoclaves. It is also possible to cure local areas without heating the entire assembly. Energy consumption for curing is 1·2 to 3 kW/m².

Induction heating is rapidly finding its place in bonding processes, especially where local heating, surface heating, extremely rapid heating, and extremely high temperatures are required [12]. For example, 'hot-shot' bonding is being used to reduce the handling time of room-temperature-curing adhesives. This means heating at high temperatures for short periods of time of spots along the bond line with a time cycle approaching that of spot welding. These spots hold the assembly together to permit immediate removal from the jig.

High-frequency dielectrical heating has also been investigated as a method by which the adhesive is heated before the metallic parts.

Another new method of curing uses ultrasonics [13]. Mechanical vibrations from an ultrasonic horn are transmitted to the adhesive layer. Friction and viscoelasticity produce heat which melts and cures the adhesive.

Pressure is required during curing to keep parts in alignment and to overcome distortion and thermal expansion in the adherends. To keep the bond line thickness required, it is recommended that pressure be applied to adhesives that do not require it also; this improves wetting, the adhesive penetrates into the surface irregularities and the parts are kept in alignment.

In certain types of joints and in non-critical joints, bonding can be effected without the use of clamping tools and fixtures; surfaces to be bonded need only to be in contact and held in place while the bond sets. In other cases, elaborate jigs and fixtures are needed. It is essential that the jigs and fixtures are easily operated and reliable. The exact positioning of the parts should be precise. It is also important that the clamping pressure is uniformly distributed over the surfaces to be bonded. And if the parts, as occurs very often in actual production, do not fit well, a certain additional external pressure is required in order to deform both parts sufficiently until they will fit perfectly. However, excessive pressure should be avoided to prevent too much adhesive being squeezed out so that the joint is starved. Figure 4.16 shows diagrammatically different possibilities for clamping and distributing the pressure. Mechanical pressure does not provide the uniform pressure needed if the parts have curved surfaces; in such cases curing in an autoclave is more suitable.

When bonding large structures expenditure may markedly be reduced

PROCESSES FOR ADHESIVE BONDING

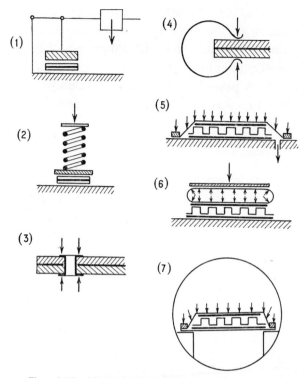

Fig. 4.16 Methods of applying pressure to the assembly while being bonded: (1) lever press; (2) spring press; (3) hollow rivet (it can be removed after the curing, if necessary); (4) clamps; (5) vacuum bag method; (6) sealed pressure bag method; (7) autoclave method.

if the adherends are fixed by spot welding, 'hot-shot' bonding, or applying drops of cyanocrylate-based adhesives.

To avoid the sticking of the parts to the jigs and fixtures, the latter should be coated with silicone laquers or oils. Any excess adhesive should be removed or cleaned up before the curing process has advanced with acetone, ethyl methyl ketone or similar solvents. The cured adhesive is virtually impossible to remove except by machining.

The temperatures specified for curing are always bond line temperatures. For this reason extra time must be allowed for the heat to penetrate to the bond line, so that the latter attains the curing temperature. The temperature of the bond line is usually measured with thermocouples positioned in different areas of the bonded assembly. The time to heat the tools and fixtures takes up often a large part of the bonding cycle.

It is important in all cases to allow the assemblies to cool gradually after completion of the cure. This will eliminate to a great extent the development of stresses in the joint, which may cause deformation. This danger is particularly possible in joints between thin parts and between parts with different coefficients of thermal expansion.

4.4 Additional processing

After curing, the bonded assemblies may be ground, filed, drilled, tapped and machined. However, the feed should always be small, the tools used should be sharp, and machining should be performed at high cutting speeds.

5
Mechanical Properties and Performance of Adhesive Bonds

5.1 Bond strength

The strength of a bond is expressed by the load per unit area of joint. The mean strength values obtained when bonding with room-temperature-curing adhesives are 1 to 2×10^7 N/m², and 2 to 6×10^7 N/m² with heat-curing adhesives.

The proper strength (cohesion) of the adhesive and the sticking (adhesion) between the adhesive and the metal are conditions upon which the bond strength depends. Whereas the proper strength of an adhesive is a constant and is relatively low, the adhesion with the different materials varies and determines the bond strength. The latter is influenced by a great number of factors.

The following are some of the factors that vary with the adhesive used: molecular structure and polarity, components, wetting properties, rheological processes, and curing processes.

Factors that are dependent on the adherends are: molecular structure and polarity of surfaces, wettability, surface finish, material and strength properties, and thickness.

Factors that are dependent on the bonding processes are: application of adhesive, atmosphere in the working rooms, curing conditions (pressure, temperature, duration), type of joint and bond line thickness.

Factors that are dependent on the end-service conditions are: type and rate of loading, and environment (weathering, temperature, corrosive media).

The effect of individual factors cannot always and unequivocally be distinguished since the factors are interdependent to a varying degree, being sometimes superimposed. Thus, both the preliminary determination of the most favourable conditions for bonding and the evaluation of the results obtained are made very difficult.

5.2 Stress distribution in bonded joints

Results from numerous studies have established that the stresses induced by external forces are not uniformly distributed over the area of bonding. As a result of the differences in elastic properties of the adhesives and adherends,* there is a concentration of stresses at the ends

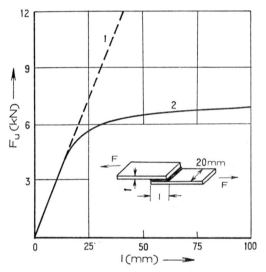

Fig. 5.1 Effect of the lap length (l) on the failure load (F) for simple lap joints (adherends: aluminium alloy; adhesive: room-temperature-curing two-part epoxide): 1 – ideal lap joint; 2 – real lap joint.

of the joint. Since these stress peaks most often exceed several times the stresses determined analytically, they are after all those that are decisive for the strength of the bond. It is because of these stress concentrations that failure load cannot be accepted as a generally valid criterion for the

* It should be borne in mind that the strength of adhesives is considerably lower than that of the metals to be bonded. The modulus of elasticity of structural adhesives is 0.03 to 0.06×10^{11} N/m², while that of steels is 1.5 to 2.5×10^{11} N/m². In many cases the coefficients of thermal expansion also differ considerably.

MECHANICAL PROPERTIES AND PERFORMANCE 57

quality of a bonded joint. Even if for a specified joint the ultimate strength τ_u may be expressed by the ratio

$$\tau_u = F/A$$

where F is the failure load and A the bond area, the existence of a relationship 'failure forces ratio=bond areas ratio' cannot be assumed for another joint type. This results from the fact that the strength of a joint is only proportional to the width of the joint, and is not proportional to its length. As seen from the example in Figure 5.1, the strength increases with an increase in the length only up to a limiting value. In this case, the assumption that the stresses are uniform is valid only for $l < 14$ mm. The maximum strength is attained at $l \approx 50$ mm, and any further extension of the lap length would serve only to satisfy manufacturing tolerances.

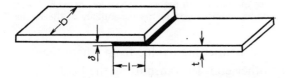

Fig. 5.2 Simple lap joint: b – adherend breadth; t – adherend thickness; l – lap length; δ – bond line thickness.

Simple lap joints (Figure 5.2) are by far the most widely used adhesive joints. Owing to this, further discussion will concern in general lines this type of joint.

Simple lap joints have been the subject of considerable analysis. The purpose of the analytical and experimental investigations has been to establish the distribution of stresses in the bond line depending on the mechanical and processing properties of the adherends and the adhesive on one hand, and the joint geometry on the other. It has been found, in principle, that the stress peaks at both ends of the lap of a simple lap joint are caused by:

1 The different strain in the adhesive and the adherends under load.
2 The bending moments due to the offset of the load from the centre line of the adherends.

This is illustrated in Figure 5.3. Assuming the validity of Hooke's law when the strips are loaded by forces F, the resulting extension strain

E

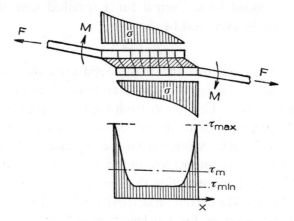

Fig. 5.3 Non-uniform stress distribution along the bond area of a simple lap joint under load (exaggerated): σ – tensile stress distribution in the adherends; τ – shear stress distribution in the adhesive layer; $M = F(t + \delta)$ – bending moment due to the eccentricity of the forces F.

decreases from a distinct high value from where they enter the overlap to zero at the other end. In transmitting the load from one strip to the other, the adhesive layer is deformed due to non-linear strains in the strips. The deformations (the stresses, respectively) in the central area of the bond are negligible, while they reach maximum values at the ends of the joint. Figure 5.4 shows a typical broken simple lap shear test specimen; it is seen that failure in the bond line occurs at the end edges of the strips where the stress is τ_{max}.

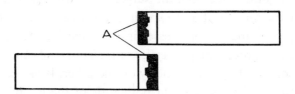

Fig. 5.4 Broken simple lap shear specimen: A – adhesive remainders.

In an early theoretical investigation, Volkersen [1] established that with certain assumptions, such as identical adherends with equal thickness t, equal modulus of elasticity E and the validity of Hooke's law, it is possible to determine the relationship between adhesive layer displacements and adherend elongation. Proceeding from a second-order linear differential equation he obtained the equation:

$$\tau(x) = \tau_m \frac{\sqrt{\tfrac{1}{2}\Delta}}{\sinh\sqrt{2\Delta}} \left\{ \cosh\left(\frac{x}{l}\sqrt{2\Delta}\right) + \cosh\left[\sqrt{2\Delta}\left(1-\frac{x}{l}\right)\right] \right\}$$

It is seen that the non-uniform stress distribution can be expressed by the dimensionless factor Δ:

$$\Delta = \frac{Gl^2}{Et\delta}$$

where G is the shear modulus of the adhesive.

The stress concentration factor can be written as follows:

$$n = \frac{\tau_{max}}{\tau_m} = f(\Delta) = \frac{1}{k}\coth\frac{1}{k}$$

where

$$\frac{1}{k} = \sqrt{\tfrac{1}{2}\Delta} = l\sqrt{\frac{G}{Et\delta}}$$

Knowing the values included in these formulae, the strength of the joint may be calculated from:

$$F = \tau_m A = \frac{\tau_{max}}{n} lb$$

where b is the breadth of the joint.

In their studies on stress distribution in simple lap joints Golland and Reissner [2] and other scientists have considered the influence of normal (tensile) stresses in the adhesive layer caused by bending moments due to the offset of the tensile forces F from the centre line of the joint (Figure 5.3). It has been found that if the metal starts to stretch plastically before the joint fails in shear, these normal stresses will increase rapidly inducing peeling action and may cause the joint to fail. In normal use, however, when light gauge metals and normal lap lengths are used, the normal stresses are sufficiently small to be ignored and the shear strength of the adhesive is considered to be the criterion of strength.

The factor Δ is, in principle, valid for all materials, since it contains the most important constants characterising them. Its use, however, for practical calculations is not expedient. If given materials are to be bonded with a given adhesive, the G/E ratio is a fixed one. Since it is known that the bond line thickness should be the smallest possible, δ may be assumed a constant and what is left over is only the l^2/t ratio.

Proceeding from this, de Bruyne introduced for practical comparison purposes the so-called 'joint factor' f, expressed by the ratio:

$$f = \sqrt{t}/l \text{ (mm}^{-0.5}\text{)}$$

Hence, under comparable conditions and for a given adhesive, simple lap joint strengths are constant at constant values of f. The relationship between the joint factor and tensile shear strength must be determined for each different type of adhesive and adherend material by laboratory tests in which lap length and adherend thickness are varied and processing details are kept constant. A typical graph obtained in this manner is shown in Figure 5.5.

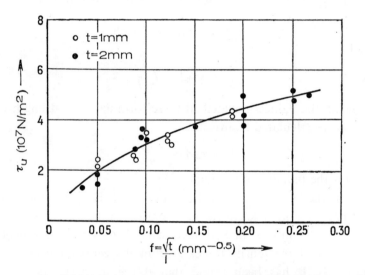

Fig. 5.5 Variation of tensile shear strength (τ) with the geometry of the joint expressed by the joint factor ($f = \sqrt{t}/l$) for simple lap joints (adherends: aluminium alloy; adhesive: heat-curing phenolic 'Redux').

Based on all that has been stated so far, German workers have shown [3] that, for a given adhesive–adherend combination, bond failure stresses are generally proportional to the square root of f:

$$\tau_m = a\sqrt{f} \pm 18 \ (\times \ 10^7 \text{ N/m}^2)$$

(t and l in mm). The normal scatter is $\pm$ 18 per cent while, for example for a phenolic adhesive (Redux 775) and AlCuMg–alloy adherends, the proportionality factor is $a = 10 \cdot 7$.

MECHANICAL PROPERTIES AND PERFORMANCE

5.3 Static strengths of bonded metal-to-metal joints

There are four basic types of loading an adhesive joint: tensile, shear, peel and cleavage (Figure 5.6).

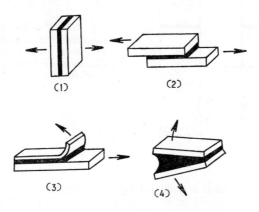

Fig. 5.6 Basic types of loading adhesive bonds: (1) tensile; (2) shear; (3) peel; (4) cleavage.

5.3.1 Direct tensile strength

Direct tensile strength of a bond is determined by the amount of tensile load per unit area it takes to break the bond when the acting forces are applied perpendicularly to the adhesive plane [Figure 5.6(1)]:

$$\sigma_t = F/A \text{ (N/m}^2\text{)}$$

Tensile strength is also dependent to a certain extent on the geometry of the joint. It should be noted that the direct tensile strength values that can be reached are considerably higher than tensile shear strength values obtained with simple lap specimens. But, nevertheless, tensile loaded joints are rarely used in practice because usually the acting forces do not remain perpendicular to the bond line, bending takes place, and the joint will fail at loads lower than laboratory tensile tests (Figure 5.7); moreover, when joining light-gauge sheets, the bond area is too small.

5.3.2 Tensile shear strength

Tensile shear strength is most widely adopted as a measure of the ultimate shear strength of an adhesive bond loaded in tension [Figure 5.6()2]. In this type of loading the forces act in the plane of the adhesive

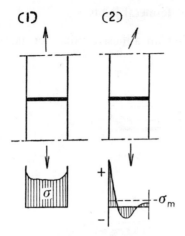

Fig. 5.7 Tensile stress distribution in butt joints: (1) under ideal loading; (2) under normal (non-axial) loading.

layer. Shear strength is expressed by the ratio of failure load to bond area:

$$\tau_u = F/A \ (\text{N/m}^2)$$

The mechanical properties of bonded joints are the highest at loads in shear. For this reason simple lap joints are the most commonly used adhesive joints and have been best studied so far.

The mean tensile shear strength of a lap joint is affected above all by the following factors, which do not depend on the adhesive: adherend material characteristics, adherend thickness, lap length and bond line thickness.

Influence of adherend material characteristics—The general conclusion can be drawn, on the basis of investigations carried out so far, that τ_u increases in an approximately linear manner with the increase in yield strength of the material of the parts to be bonded. High-strength materials (having a high yield strength) are more resistant to the action of external forces since they possess a low elongation; hence, they may be loaded to a considerably higher extent than less rigid materials before arriving at the same critical strain value in the adhesive layer that causes failure. Therefore, in the case of high-strength adherend materials, stress on the adhesive layer is eased whereas in the case of less rigid materials, it is loaded to a relatively higher degree. It is evident, that in the first case the bonded joints obtained can withstand heavier loads and

MECHANICAL PROPERTIES AND PERFORMANCE   63

the bonding efficiency factor (see page 64) has a high value, whereas, in the second case the bonded joints possess a lower specific strength and a lower bonding efficiency factor.

Influence of adherend thickness—When simple lap joints are loaded in tension, a bending moment due to the eccentricity of the forces is induced. The normal stresses resulting from this moment produce a

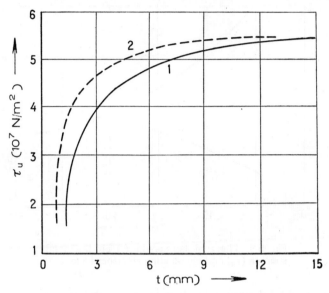

Fig. 5.8 Effect of the adherend thickness (t) on the bond strength (τ_u): 1 – for aluminium; 2 – for stainless steel.

peeling load at the ends of the joint. It follows from this that the thicker the adherends, the greater is the rigidity (the strength of the joint respectively) because the peeling effect of the bending moment is reduced. However, the structural efficiency of the thick parts is low, since, at a constant lap length l, the bond strength τ_u increases with adherend thickness t not proportionally, but only at the rate of $\sqrt{t}$ (Figure 5.8). Therefore, a double lap (or a double strap) joint should be used when bonding thick parts to improve structural efficiency.

Influence of lap length—The lap length l exerts the strongest influence on the tensile shear strength of a bonded joint. The relationship between the mean tensile shear strength of a bond and the lap length is shown in

Figure 5.9. At a constant thickness of the adherends, τ_u drops considerably with an increase in l. The exponential nature of this curve is determined above all by the non-uniform distribution of stresses along the bond line. The stress peak at the end of the overlap, the stress concentration respectively, increases with an increase in the lap length, and is manifested in a different manner when different materials are used.

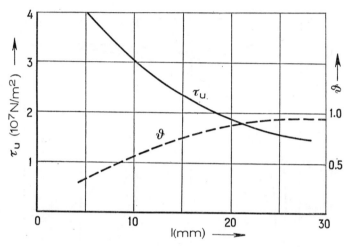

Fig. 5.9 Effect of lap length (l) on the bond strength (τ_u) and the bonding efficiency factor (ϑ).

The effect of lap length on failure load was shown in Figure 5.1. It is further seen that the increase in lap length over a certain value is not expedient because only a minimum increase in the load transmitted by the joint is achieved in this manner, or even a decrease in certain cases could be observed. In this connection, the optimum lap length should be determined in such a manner as to attain either the maximum load or the most efficient utilization of the materials bonded.

The bonding efficiency factor ϑ (Figure 5.9) is the ratio of the shear strength of the joint to the ultimate strength of the adherend material, or to its 0·2 per cent offset yield strength $\sigma_{0\cdot 2}$. The bonding efficiency factor increases with the increase of the lap length l, reaching, in a similar manner to the failure load, a maximum value, and then, in spite of an increase in l, remains almost constant. If referred to the ultimate strength of the material, ϑ could reach at most the value 1, which means that the bond strength equals that of the material. In this case, the best adhesive is the one that will provide values of ϑ closest to 1 with the least possible lap length. Or, if referred to the yield strength, ϑ could acquire values greater than 1. Since, in practice, the yield strength of the material is accepted as the load limit, in the case when $\vartheta > 1$, the degree of bond utilization is low, and, hence, a reduction in the lap length l is possible.

Influence of bond line thickness—Experimental studies have shown that an increase in bond line thickness is related to a decrease in bond strength. This relationship can be explained in part by the fact that for thinner adhesive layers the greater forces over the interfacial surfaces (adhesion forces) prevail, whereas, for thicker adhesive layers, the bond strength is determined above all by the cohesion forces in the bulk adhesive. And, as is known, adhesion strength is always higher than cohesion strength.

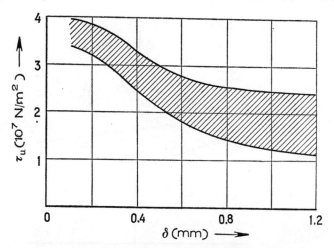

Fig. 5.10 Effect of the bond line thickness (δ) on the bond strength (τ_u) for simple lap joints bonded with heat-curing adhesives.

Figure 5.10 gives results of tests showing the influence of bond line thickness on the tensile shear strength. Generally, there is a considerable drop in strength in adhesive layers with a thickness greater than 0·1 mm, while it is approximately constant in layers greater than 0·5 mm in thickness. This leads to the conclusion that in the case of a bond line thickness greater than 0·5 mm the cohesion forces in the adhesive determine the bond strength.

On the one hand, surface roughness of the adherends limits in practice the bond line thickness; they should not come so close as to produce metal-to-metal contact in order not to break the adhesive layer. The physical properties of the adhesive layer also influence the thickness since the quality of the bonds is lowered at thicknesses greater than 0·2 mm when the capillary action is lacking (the adhesive does not fill the gap). It has been experimentally established that, in most cases, the most adequate bond line thickness falls in the 0·05 to 0·15 mm range.

5.3.3 Strength under action of combined external tensile and shearing loads
The influence of the combined action of shearing and normal stresses on the bond strength could also be of interest. Some results of tests carried out to explain this influence are indicated in Figure 5.11. It appears that in most cases the different combinations of shear and normal stresses do

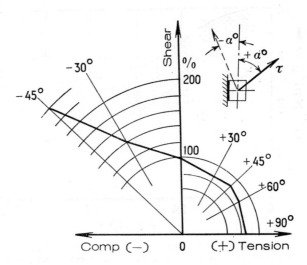

Fig. 5.11 Strength of bonded joints subjected to different combinations of shear (τ) and normal (σ) stresses ($\tau = 100$ per cent for pure shear).

not lead to strength characteristics lower than those obtained in shear or direct tensile testing. Owing to this, the latter are sufficient to evaluate the adhesive. External compression loading increases considerably shear strength. This is the case found in practice when bonding friction lining materials to brake shoes and clutch plates.

5.3.4 Peel strength

Adhesive bonded joints are particularly sensitive to peeling load [Figure 5.6(3)]. This type of load does not act on a given area; failure of the adhesive layer occurs in a progressive fashion along a line, perpendicular to the axis of loading. For this reason, the value of peel strength is expressed as the ratio of the failure load to the breadth of the bonded area:

$$\sigma' = F/b \text{ (N/m)}$$

Data for peel strength are not directly valid for joints of practical design.

In design practice, peel is considered only in so far as measures are taken to prevent the development of peeling forces (see Figure 7.1). Peel tests, however, have received wide acceptance as quality control tests. They are also a valuable process inspection tool, and can be used for qualifying adhesives, as well as studying the distribution of bonding forces in a joint.

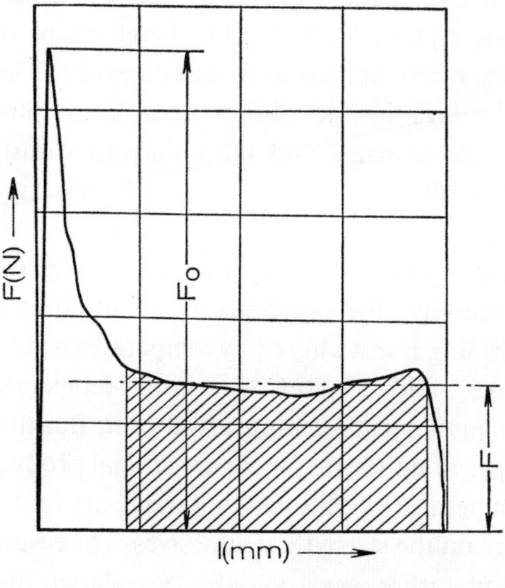

Fig. 5.12 Typical peel diagram showing the variation of the peel force (F) with the length (l) of the bond.

The magnitude of the resistance of the adhesive to peeling is measured and recorded by the weighing mechanism of the testing machine. Curves are obtained similar to the one shown in Figure 5.12. It is seen that a relatively high load F_0 is initially needed to start failure; then the load F, required to continue peeling, remains fairly constant. The peel strength is usually defined as the average value of the peeling load curve, excluding the first and last portions of the test.

The resistance to peel is almost independent of bond line thickness; this is contrary to results obtained from other tests. Peel strength increases with the increase in the modulus of elasticity and in the thickness of the adherends. The peel resistance is not simply determined by the rigidity or softness of the adhesive layer. There are some indefinable properties which determine peel behaviour for any given set of adherends. For example, plasticizing an epoxide resin with an ester-type plasticizer does not have the slightest effect on the peel performance of the adhesive, although it will greatly improve the impact strength of the compound, and therefore, the adhesive layer will be capable of absorbing external stress. On the other hand, the use of a reinforcing fabric in the same epoxide resin layer will increase five- to sixfold the peel strength of the same system.

In cleavage loading [Figure 5.6(4)] severe localized loading occurs on one side of the joint only, while the other side is under no load at all. This type of load should, therefore, be avoided. Cleavage strength is very rarely quoted in reference works. The usefulness of such information lays in the ability to indicate how an adhesive would resist service damage, and the ability to withstand further processing after assembly.

5.3.5 *Influence of temperature on the strength of bonded joints*

Generally, the performance of existing adhesives for metal-to-metal bonding is not altered by temperatures up to 100–120°C for heat-curing types, and 50–80°C for room-temperature-curing types, the bond strength declining abruptly after it. But this limit varies for the different types of adhesives; there are special products resistant to extremely high temperatures. This makes it necessary to determine the effect of temperature on the strength of adhesives. The results of numerous tests, in which temperature was a variable, are shown in Figure 5.13.

Most adhesives are resistant to the low temperatures mostly used in general engineering practice (to about −50°C). Resistance of adhesive bonds to lower temperatures is of interest for special cases only, and the bonds should be capable of withstanding the embrittling effects of these temperatures.

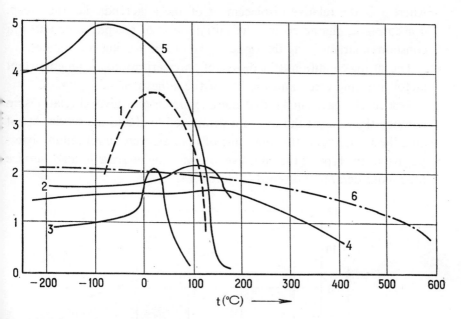

Fig. 5.13 Effect on the temperature (*t*) on the tensile shear strength (τ) for different types of adhesives: 1 – phenolic-polyvinyl formal; 2 – heat-curing epoxide; 3 – room-temperature-curing epoxide; 4 – epoxide-phenolic; 5 – epoxide-nylon; 6 – polybenzimidazole. Adherends: aluminium alloy for 1 to 5; stainless steel for 6.

5.3.6 Creep strength

The strength of an adhesive bond subjected to high loads for a considerable length of time is markedly affected by the physical properties of the adhesive. As structural adhesives are plastics materials, their behaviour should be taken into consideration in the regions of both elastic and plastic deformation. It is known that in the case of prolonged loading, plastics are apt to creep to a higher or lower degree. This is the reason why creep is to be expected in the adhesive layer of metal-to-metal bonds. It is obvious that this phenomenon could exert a decisive influence on the bond strength, particularly in cases when a rise in temperature is also added to external loading.

It has been established [4] that the relationship of the elongation of specimens tested for creep, versus time, plotted on a log-log chart, is practically linear and can be expressed as:

$$y = at^n$$

where y is the relative displacement of the adherends, i.e. the creep, a a constant, characterising the initial value of creep, t time and n a constant, characterising the slope of the plot in log-log coordinates.

The curing conditions (the degree of polymerization) are an important factor affecting creep and the life of the adhesives.

Test results have shown that there exists an unequivocal relationship between the value of the constant prolonged load and the service life of an adhesive joint, i.e. the total time of load exposure until failure, irrespective of the type of the adhesive, or, with the increase in load, service life is considerably shortened (Figure 5.14).

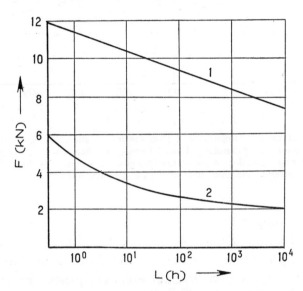

Fig. 5.14 Effect of the constant load value (F) on service life (L) of simple lap tensile shear specimens (adherends: aluminium alloy): 1 – for heat-curing phenolic adhesive; 2 – for room-temperature-curing epoxide adhesive.

Results obtained from creep tests with full-scale production assemblies (bonded sheet-metal beams, for example) indicate that creep in these cases follows the same time law observed in small specimens.

On the basis of experiments carried out so far for periods that are of significance to engineering practice, values equal to 0·25 to 0·4 of the tensile shear strength could be accepted as guide lines for long-term strength.

5.4 Dynamic strength of bonded metal-to-metal joints

The fatigue resistance of adhesive bonded joints is one of their primary assets. To establish it there is need for studies of a long duration. Figure 5.15 shows some results obtained in a comparative investigation into the strength of solid, riveted and bonded test specimens loaded in pulsating tension. It is seen that the position of the curve representing fatigue

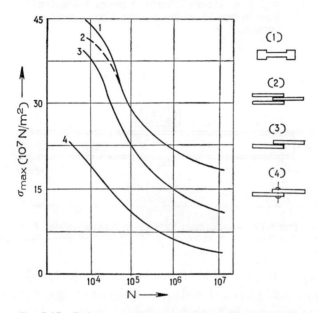

Fig. 5.15 Fatigue strengths of aluminium-alloy specimens under pulsating tensile load: σ_{max} – maximum tensile stress in the metal; N – number of load cycles; 1 – for standard solid specimens; 2 – for bonded double-lap specimens; 3 – for bonded simple-lap specimens; 4 – for riveted specimens.

behaviour of the simple lap-bonded joint lies between the curves for the metal itself (solid specimens) and the riveted specimens. The curve for double lap-bonded joints is almost identical to the curve for the metal.

The results obtained in East Germany from tests on full-scale structural members fabricated by welding, riveting and bonding [5] are also convincing. I-beams with a length of 2 m, made of AlCuMg-alloy sheets were used. The cross-section of the beams (Figure 5.16) was selected in

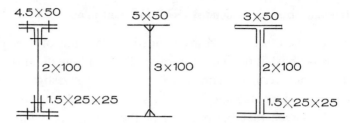

Fig. 5.16 Cross-section of Ī-beams (all dimensions in mm). From left to right: riveted; welded; bonded.

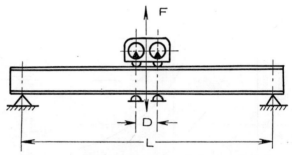

Fig. 5.17 Arrangement for Ī-beam fatigue testing: $L = 1,900$ mm; $D = 150$ mm.

such a way, as to provide in all three cases an equal moment of resistance. Figure 5.17 gives a sketch of the fatigue testing condition. Results of the tests are given in Table 5.1.

When loading bonded test specimens in pulsating tensile shear, fatigue failure occurs in many cases in the metal (especially with light

TABLE 5.1 Results of tests performed on Ī-beams

Type of beam as in Fig. 5.16	Riveted (1)	Welded (2)	Bonded (3)
Moment of resistance, M_r cm^3	30	30	30
Area of cross-section, A—cm^2	9.5	8	8
Deflection, δ mm, under static bending load $F = 15$ kN when length between supports $l = 800$ mm	2.24	1.85	1.98
Fatigue strength, σ_e (10^7 N/m^2), under reversed bending load ($N = 10^7$ cycles of load)	4.5	5.5	10.5
Number of load cycles N when failure occurs under reversed bending stress $\sigma_b = \pm 12 \times 10^7$ N/m^2	100,000	120,000	630,000

MECHANICAL PROPERTIES AND PERFORMANCE

metals). Therefore, the upper limit of the strength value, i.e. the number of loading cycles required to cause failure, is not attained. Owing to this it is accepted that fatigue strength should be related to a prescribed period of time, i.e. a specified number of loading cycles.

For simple lap bonds subjected to pulsating shear load, similar rules with respect to the relationship between strength and joint geometry are valid as in the case of static load. However, the statically optimal lap length differs from the dynamically optimal lap length. An increase in the length of lap in simple lap joints leads to an increase in fatigue life. The latter, however, decreases slightly with an increase in thickness of the adherends when the l/t ratio is kept constant. It is still not clear whether the joint factor is valid in the case of dynamic loading.

Impact testing is a special field of dynamic investigations, characterized by high speeds of loading and deformation. Impact strength data has its usefulness in indicating embrittlement.

5.5 Ageing and resistance to environment conditions

In most structural adhesives a natural ageing is observed, i.e. the strength of test specimens stored at normal atmospheric conditions, falls after a certain period of time. This fact should be taken into account in designing, and only the end-service strength values should be used.

The resistance of bonded metal joints to water (distilled, tap, salt) and some chemical reagents (petrol, different oils, acids, etc.) is of great importance for engineering practices. In these cases, conditions are greatly complicated, since in addition the bonded metal parts are often exposed to corrosion.

Moisture is also of great importance because it can markedly reduce the long-term performance of adhesives. For example, adhesives will fail at much lower stress in humid and mild elevated temperature environments than they would if humidity was absent. Tests have shown that failure at high humidity levels does not appear to be creep-induced but is catastrophic; unfortunately, the mechanism of failure is unknown [6].

It is obvious therefore that ageing and environmental testing is a vital tool in evaluating the performance of bonded metal joints. These tests, however, and investigations of the mechanism of failure require long periods of time. The extent of degradation is determined by changes in strength properties as a result of exposure to test conditions. Typical results obtained with a heat-curing epoxide adhesive are shown in Figure 5.18. When evaluating the results obtained a difference should be drawn between the absolute and relative changes. Absolute changes should always be referred to the original strength, whereas in the relative ones changes resulting from natural ageing should be taken into account.

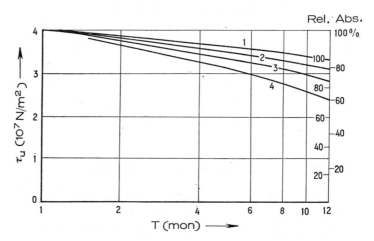

Fig. 5.18 Loss of strength in simple lap test specimens exposed to different media (adherends: aluminium alloy); T – duration of exposure (months); 1 – conditioned in closed room; 2 – exposed to weathering; 3 – immersed in tap water; 4 – immersed in sea water.

Thus, in Figure 5.18 the absolute reduction in shear strength resulting from continuous immersion in tap water for 12 months is 27 per cent; at the same time, an absolute loss in strength amounting to 15 per cent, due to natural ageing, appears. Actually, the difference (the relative reduction) between natural ageing and changes caused by the effect of tap water is 12 per cent only. It follows from this, that an adhesive bond could be considered as resistant to a given medium whenever its relative change in strength is less than or equal to 0 per cent compared with natural ageing.

In some cases, it is possible to increase bonded joint resistance to environment conditions by protective laquer coatings.

5.6 Some possibilities for increasing the strength of bonded metal-to-metal joints

5.6.1 *By construction methods*

It has been established that by changing the geometry of bonded joints, in certain cases it is possible to increase considerably their strength.

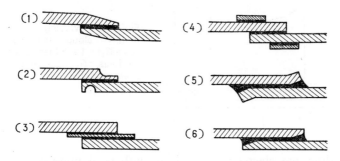

Fig. 5.19 Methods for improving the efficiency of simple lap joints.

Possibilities for this, valid for simple lap joints, are shown in Figure 5.19:

1 Bevelling of the lap ends.
2 Reducing stiffness of lap ends by removing material.
3 Inserting thin flexible plies.
4 Bonding doublers at bending points.
5 Bending of the lap ends.
6 Internal bevelling of the lap ends [7].

For example, method (5) results in a 15 per cent increase in strength under static load, and a 25 per cent increase under dynamic load.

Use of glass-fabric carriers in the bond line, e.g. a single layer of open scrim or plain-weave glasscloth, considerably improves peel strength in particular. This method provides better strength retention above the deflection temperature of the adhesive. Likewise, the effect of stresses resulting from bonding materials of dissimilar coefficients of expansion is reduced to a minimum. Particularly good results are obtained when heat-curing epoxide-based adhesives are used, whereas a certain decrease in strength may occur with phenolic adhesives. Generally, it is expedient to use dense fabrics with room-temperature-curing adhesives.

Tests have shown that the strength of simple lap joints may be increased by about 20 per cent by varying the adhesive rigidity along the lap length, so that a flexible adhesive carries the strain at the joint edges, while a brittle adhesive is used in the centre of the lap (Figure 5.20) [6].

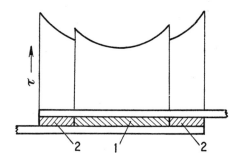

Fig. 5.20 Varying the adhesive rigidity along the lap length ($G_2 > G_1$): 1 — adhesive with shear modulus G_1; 2 — adhesive with shear modulus G_2.

Extending the elastic region of the adherend materials is another method of increasing the bond strength. The adherends are subjected to a plastic deformation prior to bonding. The deformation may be effected either in the whole body, or only in the zone of maximum deformation of the bond area. This is achieved in the following manner:

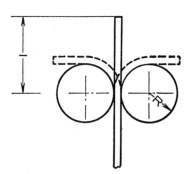

Fig. 5.21 Effecting a plastic deformation in specimen strips by multiple back-and-forth bending through 180° ($R = 10$ mm).

(1) by previous loading in tension, and (2) by repeated bending around rolls, as illustrated in Figure 5.21. It should be borne in mind that only room-temperature-curing adhesives are to be used in bonding non-ferrous metals and alloys, having a low recrystallization temperature, after the latter have been plastically deformed previously. This is necessary to avoid annealing the bonded parts, which in turn eliminates the effect of the preliminary plastic deformation.

5.6.2 By irradiation

It is known that gamma-rays influence the polymer structure of plastics to a considerable degree. An increase in rigidity, temperature resistance and in the modulus of elasticity is the result of this. Since structural adhesives are plastics materials, exposure of bonded metal-to-metal joints to radiation also causes changes in their strength. It has been established that these changes are greater if the bonds are exposed to radiation during the cure, than after it. For example, an increase in strength of about 4 per cent has been observed [8] in bonds made with a one-part heat-curing epoxide adhesive. Likewise, a certain increase in strength at elevated temperatures has been shown by some adhesives. However, more systematic studies are needed to determine the relationship between the irradiation rate and the strength at different temperatures, and in this connection, the profitableness of irradiation.

5.6.3 By mechanical vibration

The possibility has been established of increasing the strength of bonded joints by employing ultrasonic vibrations [9], and at the same time, of minimising the scatter of strength obtained during tests. The joints are subjected to the action of ultrasonic vibrations for a short period prior to curing of the adhesive. For example, aluminium strips bonded with a heat-curing adhesive, have been exposed to ultrasonic vibration for 1 min after adhesive application and assembly. After the curing of the adhesive, the specimens were tested for strength, and a failure force of about 1,025 daN was measured, the scatter being $\pm$ 40 daN. The strength of a similar joint, not subjected to vibration, was about 850 daN at a scatter of $\pm$ 150 daN; hence, the increase in strength is more than 20 per cent.

6
Testing and Inspection of Adhesive Bonds

The rapid expansion of adhesive bonding into many applications has called for investigation and testing of bonded joints in many research centres, institutes, laboratories and production plants. A large number of testing techniques used in the testing of metals have been adopted. But they had to be modified in accordance with the requirements valid for the testing of bonded joints. Owing to the fact that specific strength values obtained cannot be readily translated to other geometries or areas, poor reproducibility resulted. For this reason, work is being done to develop universally acceptable methods of testing adhesives and adhesive bonds, and standardising testing methods and conditions. A number of standard tests for adhesive bonds have already been defined in different countries and there are military specifications for specialized testing of structural adhesives (see Appendix 2).

6.1 Destructive methods for testing bonded metal joints

6.1.1 Static strength tests

Complex forces are involved in all the basic types of strength testing. A simple tensile test produces shear in addition to tensional stresses, and shear and cleavage tests involve tensional, compressional and shear stresses. In practice, however, in studying the static strength of a bonded metal-to-metal joint the prime interest lies in the normal and the shear stresses. Depending on the type of joint, the peeling load could be of importance. Behaviour under continuous stress and different environmental conditions is also of interest.

Destructive tests are performed on standard tensile testers, equipped with suitable self-aligning grips and jaws. The failure load should be so selected as to fall between 15 and 85 per cent of the full-scale capacity.

The speed of loading could be as high as 20 mm/min. The atmosphere in the room should be normal, i.e. the air temperature should be 20±5°C and the relative humidity 50 to 70 per cent. Five to ten similar specimens should always be tested, since all adhesive results are subject to scattering. Before testing, the finished specimens should be conditioned in a room with normal atmosphere. This period of conditioning should be 24 to 48 hr after the end of curing for heat-cured adhesives, and 7 days after bonding for cold-setting adhesives. The nature and amount of the failure (such as cohesion in adhesive or adherend, or adhesion) should be recorded for each specimen.

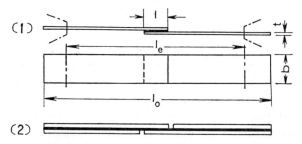

Fig. 6.1 Tensile shear test specimens: 1 – simple lap joint; 2 – two-ply lap joint.

Tensile shear testing—This test utilizes simple lap jointed specimens which are used to measure the ultimate shear stress of the bond by its resistance to a tensile load. Their form is shown in Figure 6.1(1), and the sizes according to some standards and specifications are given in Table 6.1.

To measure the strength obtained when bonding over a large area, so-called two-ply joints are employed. Strips are cut from the bonded area, and at the middle, two saw cuts are made through each adherend on opposite sides of the joint, as shown in Figure 6.1(2).

Standard tensile testers with a load range of 0 to 10 kN are used. The load at failure of the bonded joint is recorded. The tensile shear strength of the bond is calculated as the failure load divided by the shear area:

$$\tau_u = F/lb \ (N/m^2)$$

This test is widely used to qualify metal adhesives, determine the influence of different processing variables, evaluate the reduction of

TABLE 6.1 Sizes of bonded simple-lap metal test specimens

Country	Standard or specification	Sizes (mm)				
		t	b	l	l_0	l_e
USA	Fed. Std. Test Method No. 1033.1–T (ASTM D 1002–53T)	1·62 (0·064″)	25·4 (1″)	12·7 (0·5″)	228·6 (9″)	63·5 (2·5″)
USA	Spec. MIL–A–5090D	Alumin.: 1·62 (0·063″) Steel: 1·20 (0·050″)	25·4 (1″)	12·7 (0·5″)	190·5 (7·5″)	—
USA	Spec. MIL–A–14042A (Ord)	3·2 (0·125″)	25·4 (1″)	12·7 (0·5″)	165·1 (6·5″)	63·5 (2·5″)
West Germany	Std. DIN 53,281	1·5	25	12	212	100
East Germany	Std. TGL 14; 910	1·5	25	10	190	100
Czechoslovakia	Std. CSN 66 8510	1·62	25·4	12·7	194	·

strength resulting from detrimental environments, and, in general, to provide designers and engineers with useable data.

Direct tensile testing—In several cases it is necessary to determine the tensile strength of a bond when the acting forces are applied perpendicularly to the adhesive plane, i.e. when the stresses in the bond line are normal. Direct tensile strength is probably a more accurate measure of the work of adhesion than the simple lap tensile shear strength. However, there are certain difficulties linked with test procedure, the main one being the dependence of the tensile value on the geometry of the joint.

Spool-type test specimens are normally used (Figure 6.2). The ends of the specimens must be held in special pin-type grips so that the loading is axial. Standard tensile testers are used for loading until failure occurs. The tensile strength of the bond is calculated as the failure load divided by the bond area:

$$\sigma_{u_t} = \frac{F}{\pi d^2/4} \quad (\text{N/m}^2)$$

To minimize the influence of edge zones, the diameter of bond area should be greater than 15 mm. The specimens may be re-used after testing if resurfaced by grinding flat and parallel areas that contain the adhesive.

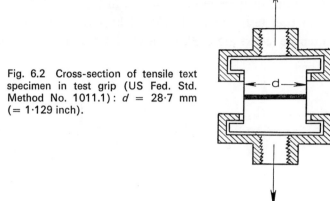

Fig. 6.2 Cross-section of tensile text specimen in test grip (US Fed. Std. Method No. 1011.1): $d = 28\cdot 7$ mm ($= 1\cdot 129$ inch).

Torsion shear testing—In certain cases it is of interest to obtain and determine pure shear stresses. To this end, torsion testing of butt-bonded ring-type specimens is most suitable. Testing is done on standard torsion testing machines, but special grips to hold the specimens are needed. The latter are loaded to torsion until bond line failure occurs and the torque N at failure is recorded. Torsion shear strength of the bond is calculated as the torque at failure divided by the polar moment of resistance:

$$\tau_{uq} = \frac{N}{M_{r_{pol}}} = \frac{N}{\frac{\pi}{16} \cdot \frac{d_o^2 - d_i^2}{d_o}} \quad (\text{N/m}^2)$$

where d_o and d_i are the outer and inner diameters of the bonded area, respectively.

Testing under combined tensile and shear load—Flat and cylindrical specimens, as seen in Figure 6.3, may be used instead of scarfed ones in this testing so that the influence of adherend rigidity could be avoided, and the phenomenon studied in pure form. They are subjected to tension on standard tensile testers.

Peel testing—Unlike the methods described, where mean strength values relevant to the whole bond area are obtained, peel testing tears or cleaves the adhesive film in a progressive fashion. A large number of different types of peel tests have been developed. Easier illustrated than described, various of these tests are shown in Figure 6.4.

The drum peel test (*a*) was developed by CIBA (A.R.L.) Limited. Standard tensile testers are used, but a special peel jig is needed.

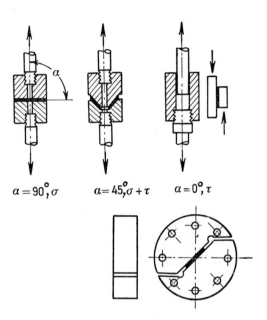

Fig. 6.3 Test specimens for determining the combined effect of shear (τ) and normal (σ) stresses on bond strength.

The climbing drum test (*b*) was developed in the USA (US Fed. Std. Method No. 1042–T). When the lower machine jaw is drawn down, the cables round the drum, by virtue of the different radii, exert a torque which makes the drum climb up the rigid plate rolling the thin sheet round it.

Method (*c*) is theoretically the most expedient, but its practical application is linked with difficulties.

The T-peel test (*d*) derives its name from the geometric form which the specimen assumes as it is being peeled. This method is easy to realize and standard tensile testers are used.

The SAAB–test (*e*) has been developed in Sweden. Its practical use is linked with difficulties.

Method (*f*) (US Fed.Std.Method No. 1041.1) is used mainly to rate the flexibility of adhesives.

Methods (*g*) and (*h*) are used to measure the peel strength of bonded specimens subjected to static or dynamic bending loads.

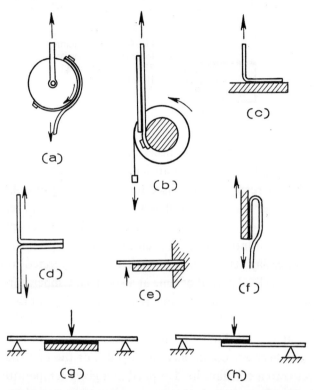

Fig. 6.4 Diagrammatic representation of different types of peel tests.

The magnitude of the resistance of the adhesive to peeling is measured and recorded by the weighing mechanism of the testing machine. The results obtained depend on the material of the test specimens (the yield strength and thickness), as well as on the test method. The results obtained with the different types of peel tests vary sometimes considerably. This makes it necessary to indicate always the method of testing used.

Peel strength is determined by the mean value of the peeling load curve, excluding the first and last portions of the test:

$$\sigma' = F/b \text{ (N/m)}$$

The cleavage strength of the adhesive can be determined similarly by using the specimen shown in Figure 6.5, conforming to US.Fed.Std. Method No. 1071–T. The grips used to hold the specimen must be of the self-aligning type.

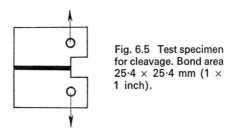

Fig. 6.5 Test specimen for cleavage. Bond area 25·4 × 25·4 mm (1 × 1 inch).

Creep testing—Creep testing is necessary because strengths determined by short-time tests do not give an indication of behaviour under continuous stress. Creep machines used for creep testing of adhesive bonds consist of a steel frame, a cantilever loading device and lead weights. In recent years, however, more satisfactory and compact spring loaded devices have been introduced. In each case the test specimens are of the familiar simple-lap tensile shear type. They are subjected to a constant load for a prescribed period of time at a specified temperature. Accurate gauges are placed on the specimens for measuring strain and creep. Temperature is a crucial factor in creep, and tests are therefore generally performed at several different temperatures.

Two creep tests are commonly employed. For the first the specimens are subjected to a constant load applied at a given temperature, and the time to failure is measured. For the second test the specimens are subjected to a constant load and the mean stress which the joint may carry for a given length of time is determined. Of secondary interest is the increase of permanent deformation of the adhesive, i.e. slip of the joint, with time. The duration of testing should be at least 1,000 hr.

6.1.2 Dynamic strength tests

Fatigue testing—Fatigue testing consists of repeated application of a given load or deformation on an adhesive system. It measures the ability of a bonded specimen to resist failure, but does not measure mechanical properties. Several strength test specimens of the types previously discussed, such as the simple-lap tensile shear specimen, may be used,

TESTING AND INSPECTION 85

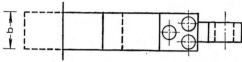

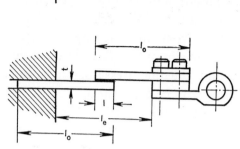

Fig. 6.6 Specimen and mounting for fatigue test in pulsating tension: $t - 1·62$ mm (0·064 inch); $b - 25·4$ mm (1 inch); $l - 12·7$ mm (0·5 inch); $l_0 - 63·5$ mm (2·5 inches); $l_e - 63·5$ mm (2·5 inches).

provided that the initial introduction of stress will not cause failure. A test specimen and mounting in conformity with US Fed.Std.Method No. 1061 are shown in Figure 6.6. The testing machines should be capable of applying a cyclic load at a rate of 1,000 to 3,600 cycles per

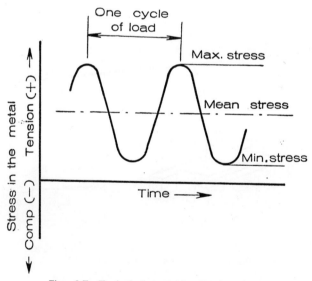

Fig. 6.7 Typical fatigue loading cycle in diagram form.

minute. Load is applied sinusoidally (Figure 6.7). A specific mean load (which may be zero) and an alternating load are applied to a specimen; the number of cycles required to cause failure, i.e. the fatigue life, is recorded and the corresponding stresses are calculated. Generally, the test is repeated with identical specimens and various fluctuating loads. Loads may be applied axially, in torsion, or in flexure. Depending on the amplitude of the mean and cyclic load, net stress in the specimen may be in one direction through the loading cycle or may reverse direction.

Data from fatigue testing are often presented in a 'stress – log cycles' diagram. From this diagram the fatigue endurance limit, i.e. the maximum fluctuating stress a specimen can endure for an infinite number of cycles, can be determined.

Fatigue strength is usually defined as the maximum applied stress, which causes joint failure after a prescribed period of time, i.e. a specified number N of loading cycles. N may be 10^7, or in accordance with the West German standard DIN 53,285, $N = 2 \times 10^7$.

Impact testing—Impact testing is usually intended for determining the comparative impact value of adhesives in shear, i.e. the toughness of adhesives, but can also be applied to direct tensile and cleavage. Normal pendulum-type impact machines may be used. Impact value u is the energy U absorbed by a specimen when sheared by a single blow of a testing machine hammer, related to bond area:

$$u = U/lb \text{ (N.m/m}^2\text{)}$$

where l is the length and b the breadth.

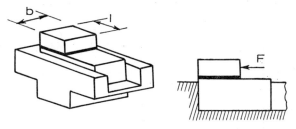

Fig. 6.8 Specimen for impact test and adapter jig for impact machines: $l = b = 25 \cdot 4$ mm (1 inch).

In the US Fed.Std.Method No. 1051 a 9·5 mm (3/8 inch) thick block of aluminium alloy, 25·4 mm (1 inch) square, is bonded to a larger block which is firmly clamped to the base of an impact testing machine (Figure 6.8). The hammer strikes the small block with a specified impact velocity and the energy absorbed in rupturing the bond is measured.

6.1.3 Permanence tests

The test methods discussed above enable the basic strength properties under static and dynamic loads to be determined. In many cases, however, tests are needed to verify the suitability of an adhesive for use in environments to which the product will be subjected during service. These tests show how ageing, temperature, weathering, and corrosive media affect the performance of adhesives. Normally, simple-lap tensile shear specimens, as shown in Figure 6.1, are used. After being exposed for a specified period of time to the appropriate conditions, the specimens are subjected to tensile shear testing. Accelerated tests that duplicate in a short time the performance to be expected for periods of months and years have been developed too.

Testing for ageing resistance—A large number of uncovered test specimens are placed in a room with normal atmosphere. After prescribed lengths of time (for example, 0, 1, 4, 13, 26, 52 weeks), five to ten specimens are tested for strength. The testing should cover a period of one year or more, but not less than six months. Based on the results obtained, diagrams are plotted showing the reduction in strength of the bond as a function of storage duration and the type of adhesive.

Short-term tests by accelerated ageing may be used, the specimens being conditioned in elevated temperatures. Care should be taken, however, to control correlation with natural ageing.

Testing for temperature effects—Simple-lap tensile shear specimens are tested for strength at elevated temperatures on standard tensile testing machines, equipped with an electric heating source (for example, tubular quartz lamps). The temperature should be controllable. Tests should be performed at one or more temperatures (for example, 20, 50, 100, 150, 200, 250°C). In all cases, the specimens should be conditioned at the testing temperature prior to testing. By plotting the failure strengths against the temperatures, diagrams are obtained that can be used for qualifying adhesives.

For testing at low temperatures the testing machines should be equipped with a device for chilling the specimens. However, testing of adhesive bonds at low temperatures is of minor importance to general engineering since most adhesives are resistant to the low temperatures normally used in practice (to about −50°C). Testing of adhesive bonds at lower temperatures is of interest only for special fields, as for example space engineering [1].

Weathering tests—A large number of test specimens, fixed on special stands, are exposed to weathering. The specimens should be placed so that free access of the atmosphere is obtained for the total bond area of each specimen. Sunlight should fall at a right angle. Access of rain and snow should also be provided, but prolonged immersion in water should be avoided. After prescribed lengths of time (for example, 0, 1, 4, 13, 26, 52 weeks) five to ten specimens are tested for strength. In this case too, testing should cover a period of one year or more. Based on the results obtained, diagrams are plotted showing the percentage loss of strength. Natural ageing, however, should always be taken into consideration.

In this case also, test duration may be shortened by exposing the specimens to cyclic laboratory conditions; for example, periodic boiling in water, and drying in hot air for short periods. Here again, the respective correlation should be controlled.

Testing for corrosion resistance—The test specimens are immersed in the respective corrosive fluid (for example, petrol, jet fuel, oil, sea water, tap water). Care should be taken in the choice of materials with respect to adherends and containers in that they are unaffected by the chemicals used in the test. The specimens must be so suspended or placed in the containers as to ensure full contact of the fluid with the bonded area. In this test again, five to ten specimens are removed from the fluid after specified lengths of time (for example 0, 1, 4, 13, 26, 52 weeks) and are tested for strength. What has been said above with respect to the duration of tests and the evaluation of the results obtained is also valid here.

Test duration may be shortened by cyclic immersion of the specimens, but correlation with continuous immersion should be controlled.

In some cases the test specimens may be subjected to conditions of extreme humidity in a test cabinet (75°C and 95–100 per cent relative humidity), e.g. for 15 days, while others may be placed for 30 days in a standard salt-spray cabinet which gives a corrosive saline fog.

Testing of cured adhesives—The basic constituent of bonded joints is the adhesive layer. Being a polymer material, its properties are entirely different from those of the metal adherends. Owing to this, the cured adhesive may be studied in detail in order to elucidate certain relationships. Yet, the results obtained are of interest to certain theoretical investigations only, since there is a general non-correlation between the mechanical properties of the polymers and how these polymers behave in adhesive joints.

6.1.4 Preparation of test specimens

Certain requirements should be met in the preparation of the test specimens, because it influences the results obtained. Fixtures should be used to provide adequate contact and to hold the specimens in alignment.

From a practical point of view, the preparation of the specimens should be easy and rapid. For this purpose, two methods that are different in principle may be followed. Each specimen is prepared separately in the first method, whereas, larger panels (often pre-slotted) are bonded in the second, and then cut into individual test specimens. The cutting operation should be done so as to avoid overheating or mechanical damage to the joints. Care should be taken that all edges of the metal specimens and panels which fall within, or which bound, the bonded joints should be machined true (without burrs or bevels and at right angles to faces) and smooth before surface pretreatment and bonding.

6.1.5 Qualification tests

Such tests are performed on the first part of a new design. The part is cut into test specimens to determine if the design and the bonding process are satisfactory.

6.2 Non-destructive inspection methods for adhesive bonds

As it would be uneconomical to destroy a ready part in order to determine its quality, the production quality control should be carried out, if possible, non-destructively. Most of the existing non-destructive tests, however, do not give a conclusive indication with regard to the strength of adhesive bonds, but most are limited to detection of voids between adherends. So, destructive testing will have to be employed until non-destructive test methods capable of predicting the performance of adhesive bonds under specific stresses are developed.

6.2.1 Acoustic inspection methods

Acoustic flaw detection holds considerable promise for the engineering industry and a number of acoustic techniques have been developed for the examination of bonded joints.

The tapping method is one of the oldest used. The inspector taps the bonded joint with a coin or a small hammer with a specified weight. The sound of the tapping over a well bonded section is different from that over an unbonded area; this difference in sounds is used to detect voids and delamination. Tapping is only a qualitative and comparative method. Its realiability is based on the sensitivity of the inspector's ear and his experience. When the complete bonded structure has a low quality, it is doubtful if this will be detected.

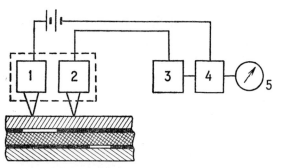

Fig. 6.9 ČIKP-2 tapping device: 1 – vibrator, tapping periodically on the surface; 2 – receiver (piezoelectric); 3 – frequency filter; 4 – amplifier; 5 – indicating instrument.

This technique, defined as the *analysis of the frequency spectrum* received from a mechanically shocked structure, has been improved by the introduction of mechanically or solenoid operated tapping devices, and microphones which pick up the sound and transmit it to a loudspeaker through an amplifier. The output of the microphone can also be fed through a frequency filter which passes only those frequencies generated when tapping over an unbonded area; the output from the filter operates an indicating instrument, lights a signalling lamp, or rings a bell. Figure 6.9 shows a device based on spectrum analysis developed in the USSR [2]. In a similar device, developed in the USA for NASA, a wire brush is moved across the surface of the object, and a scratching sound is produced which changes in frequency when differences in bonding characteristics are encountered [3]. The Eddy-Sonic method [4] is based on the principle that the structure is excited by the flow of eddy currents to generate a sound field.

Ultrasonic techniques are also used for the non-destructive inspection of adhesive bonds. Vibrational energy is transmitted into the structure to provide detection of unbonded areas. The form of wave energy utilized can be readily controlled and propagated in the structure tested. The introduction of automatic scanning and recording systems permits continuous 100 per cent inspection at comparatively high speeds. There are several basic ultrasonic flaw detection techniques [5].

The ultrasonic pulse technique consists of pulsing a transducer with a short burst of electrical energy with the resulting acoustical pulse travelling through the test material causing reflections or echoes at interfaces [6]. The pulsed-through transmission method uses two separate transducers, a transmitter and a receiver, one on each side of the component. Flaws are indicated by a reduction in amplitude of the received signal. With the one-sided pulse echo reflection method, echoes are received by a second transducer, or by the same transducer, and are processed so that their trip time and/or relative amplitude can be observed. The immersion method, with the part immersed in water, provides testing flexibility, since the transducer can be moved underwater to introduce a sound beam at any desired angle.

The impedance technique is defined [5] as the measurement of the electrical loading characteristics of an excited transducer when coupled to a material. Physical changes in the material will affect its reactive loading (wavelength) or resistive loading (attenuation) will vary the electrical impedance of an exciting transducer. Typical characteristics to be monitored include changes in resonant frequency and vibrational amplitude. Figure 6.10 shows the block diagram of an impedance system developed in the USSR [7]. When the feeler (9) is not in contact with the surface of the structure tested, the load acting on the probe, as well as the reaction force, are zero. After the probe is pressed to the surface, a reaction force is induced causing a deformation of the bottom piezo-element (4), and a corresponding rise in voltage. The greater the mechanical impedance of the structure at the point of contact with the feeler, the higher the voltage. If voids in the bond line are present, the mechanical impedance and the voltage acting on the piezoelement (4) drop abruptly. This change is recorded by the reading of the indicating instrument (6) connected to the output of the amplifier. If the voltage falls below a given value the relay (7) lights the signalling lamp (8) built in the probe. The system described works within a frequency range from 2,000 to 7,000 Hz.

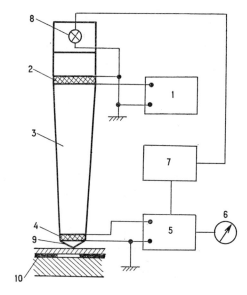

Fig. 6.10 IAD-2 acoustic impedance tester:

1 – sound generator
2 – emitting piezoelement
3 – sound conducting rod
4 – dynamometric piezoelement
5 – amplifier
6 – indicating instrument
7 – relay device
8 – signalling lamp
9 – feeler
10 – adhesive layer with disbonds.

An ultrasonic impedance instrument, the Fokker Bond Tester, is believed to be the only system at present that can give an accurate direct quantitative reading of the quality of an adhesive bond [8]. The resonance frequency and amplitude of an unbonded system are first determined by coupling the probe to a calibration specimen of the same material and thickness as the top face adherend, thus defining zero bond strength. Then the transducer is coupled to the bonded joint, thus inducing loads into the joint. The change in loading of the transducer causes changes both in the frequency at which resonance occurs and in the amplitude of vibration. Measurement of these responses takes place simultaneously. The instrument can be adjusted initially to give a trace on a cathode-ray oscilloscope with a vertical peak which moves horizontally depending on the quality of the joint (Figure 6.11). The measurements are related to resonance tests on test specimens. A close relationship has been found to exist between the tests performed on these and destructively obtained strength figures for them.

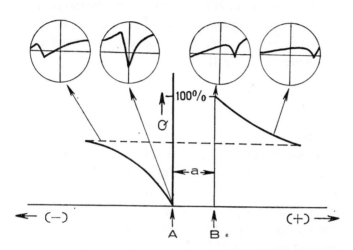

Fig. 6.11 Typical instrument frequency shift readings depending on the bond quality level (Q) and correlation curves for overlap bonds: A – reading for unbonded adherend (calibration specimen); B – reading for bond quality level $Q = 100\%$ (a – depends on adherend thickness).

Decrement techniques are defined [5] as the measurement of vibration decay or ringing pattern following the elastic pulsing of a material. The ringing envelope is a time–amplitude function. There are two types of decrement techniques. The first involves pulsing the structure as a whole (lumped system) at some natural frequency and observing the exponential decay. The second technique uses the localized reverberation effect (distributed system) of an ultrasonic pulse beamed between two interfaces of a material. Theoretically, composite ringing patterns should indicate different decrement values for bonded and unbonded areas.

Intermodulation techniques are defined [5] as the simultaneous excitation of a structure by two frequencies, one relatively low, to excite or mechanically vary the unbonded interfaces gap, and a relatively high frequency inspection signal to detect unbonded areas. Both pulsed and continuous wave approaches are possible.

Acoustic emission is a new approach to non-destructive testing based on the concept of utilizing transducer action of a flaw in a stress field [9]. This situation results in localized plastic deformation occuring in the vicinity of the flaw which, in turn, gives up acoustical energy in several different ways. The strength of adhesive bonds may be determined by measuring the acoustic emission generated at the bond line while the structure is under dynamic stress.

6.22 Radiographic inspection methods
X-ray techniques may be used only if fillers are added to the adhesive. This results from the slight weakening of the rays in the penetration through unfilled adhesives, because the density of the latter is low. Excellent results may be obtained, for example, by adding lead oxide. It is possible in this case to detect even the smallest air and gas bubbles. Conventional x-ray equipment for flaw detection is used. The thinness of the adhesive layer, however, require rays of a great length.

Radio-isotope methods—For inspecting the tightness of combined bonded and spot welded joints radioactive isotopes may be used to check the possibility of electrolyte penetrating to the bond line during subsequent anodizing in sulphuric acid [10]. The method consists in that a radioactive sodium isotope (for example, sodium-22 having a half-life of 2·5 years) is introduced into the most active electrolyte; if there are voids in the adhesive layer the electrolyte penetrates into the clearance between the adherends. The joint is then washed clean and examined with a radiometer. If there are voids, active substance is retained in them and the radiation intensity is higher. However, the application of this method in industry is limited by the danger of radiation.

6.2.3 Special inspection methods
Capacitance measurement—The joints between sheet metals may be viewed as a capacitor, and if there are voids in the bond line the capacitance changes. Tests performed to determine bond quality by capa-

citance measurement have shown that best results are obtained with adhesive layers of great thickness or if their thickness is uniform. The latter may be achieved by inserting, for example, glass cloth in the bond line and pressure is applied during the cure; in this case it is possible to examine relatively thin layers.

The vacuum method is used primarily for the inspection of bonded honeycomb sandwich panels. A cup is used, the edge of which is fitted with a rubber gasket. The cup is placed on the panel and vacuum is applied in an attempt to pull the facing sheet from the core. A micrometer gauge fitted in the apex of the cup shows any distortion of the facing sheet through lack of proper bonding (Figure 6.12).

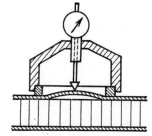

Fig. 6.12 Vacuum test cup over unbonded area.

Penetrant inspection is used for local examination of sections of seam joints. Firstly, the surface of the specimens must be cleaned and degreased. Then a penetrant solution is applied along the joint. Capillary action pulls the solution into any defect open to the surface. The penetrant on the surface is rinsed with a solvent, leaving the penetrant in the defects. Then a developer is applied to draw the penetrant back to the surface. Because the penetrants used have a brilliant colour, each defect is easy to see.

In thermal image inspection of adhesive bonded assemblies, bond discontinuities may be revealed through temperature differences of the assembly surface. The use of an ultraviolet radiation permits the direct visual detection of these discontinuities as dark regions in an otherwise bright (fluorescent) surface [11]. For practical purposes—to preclude thermal damage to the adhesive and/or heat-sensitive adherends—a phosphor is used that shows a large change in brightness with a small change in temperature in the near-room temperature range (25 to 65°C). The coatings used for this purpose provide a stable (non-settling) suspension of the phosphor in the vehicle that can be applied by conventional spray paint equipment.

Thermal infra-red inspection techniques have been used for detection of internal voids and unbonded areas in solid-propellant rocket engines and in large, panel-shaped components [12].

Cholesteric liquid crystals are compounds that go through a transition phase where they possess the flow characteristics of a liquid while retaining much of the molecular order of the crystalline solid. Since liquid crystals have the ability to reflect iridescent colours, dependent upon the temperature of their environment, they may be applied to the surfaces of bonded assemblies and used to project a visual colour picture of minute thermal gradients associated with bond discontinuities [13].

Holographic techniques—It is now possible, using stored-beam holographic techniques, to make real-time differential interferometric measurements to a precision of the order of one-millionth of a centimeter on ordinary surfaces. The holographic technique has been adapted to non-destructive testing of adhesive bonded metal structures [14]. The photographic film generally used in holographic work has been replaced by a reusable, self-developing photochromic material to reduce test time, complexity and cost. A method of stabilizing the wave-fronts at the hologram recording plane has been developed which permits reliable operation under shop vibration levels.

A simple method of inspecting bonded panels is to place them horizontally and to apply a thin sand layer on the upper surface. When the panel is vibrated, any unbonded areas will be revealed by the pattern resulting from the movement of the sand particles.

Bond quality can also be determined by making circular cuts through one adherend down to the bond line in a zone where the strength of the assembly will not be affected. The disks are pried out to expose the adhesive to visual inspection. Plugs may be inserted later in the cut-outs.

In certain cases test specimens are treated and bonded simultaneously with the production parts under the same conditions. These specimens are then tested for strength.

7
Design of Adhesive Bonded Joints

There exist many ways of making a bonded joint. Designers should, however, always design for bonding and not substitute bonding for other means of joining.

When designing parts meant for adhesive bonding, care should be taken that the stresses applied are distributed as uniformly as possible over the maximum amount of bonded area. Stresses in the adhesive should be in the direction of its greatest strength. Since most adhesives have relatively high shear strengths, it is generally best to design joints for shear stress. Direct tensile loads should be avoided, if possible, because in practice forces do not remain perpendicular to the plane of the joint; as a result cleavage stresses are developed and failure occurs at relatively low loads. It is desirable that the bond line be continuous.

High stress concentrations at the edges of the joints should be avoided. Bending stresses should be avoided, too, because they normally develop peeling or cleavage stresses.

Precautions are needed to avoid peel. The best way to do so is to prevent it from starting. Peel usually starts at edges; tapering the edges will allow a bond to conform to bending stresses. Other methods of securing ends and edges are shown in Figure 7.1.

To minimize stress concentrations on the adhesive, the latter should be less stiff than the adherends.

If the assembly is subjected to heavy impacts, bonded joints should be avoided. In special cases and if the assembly is designed to withstand vibration, good results are obtained by using an adhesive with a low modulus of elasticity and including a glass-cloth or other fabric interlayer in the bond line.

If the bonded assembly is designed to operate over a range of temperatures, the relative coefficients of expansion of the adherends and the

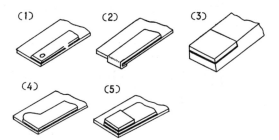

Fig. 7.1 Methods to reduce peeling stresses: (1) adding a mechanical stop (a rivet, for example); (2) beading the end; (3) recessing; (4) increasing the bond area; (5) stiffening the end by bonded doublers.

adhesive may be important. Stresses in the joint are minimized by matching the coefficient of expansion as far as possible.

Generally, designers familiar with the basic principles discussed can either design a joint around a given type of adhesive or, alternatively, find the right basic type of adhesive for any given joint design. If, for example, in certain applications, because of temperature and solvent resistance, epoxide adhesives should be employed, the joint must be so designed as to give the minimum peel effect at any time under service conditions and all the load should be transmitted, if possible, either in shear or direct tension. To illustrate the other instance, when an adhesive must have reasonable shear and tensile strengths and, at the same time, adequate peel resistance, one cannot possibly choose a thermosetting resin adhesive, but one should choose either a vinyl/phenolic, or a synthetic rubber/thermosetting resin adhesive.

7.1 Joints used in adhesive bonding

Figure 7.2 shows diagrammatically different types of joints between sheet materials, both expedient and unsuitable [1]. They will be considered briefly.

1. BUTT JOINT The stresses are perpendicular to the plane of the joint. It is inadequate for transmitting forces because the bond area is too small.

2. SIMPLE (OR UNSUPPORTED) LAP JOINT This is the simplest joint to make. It is used in many bonded structures, particularly

when thin sheet metals are being used. The strength properties of simple lap joints are discussed in detail in Chapter 5.

3 TAPERED LAP JOINT The strength is considerably higher than that of joint (2) because the bending stresses are reduced, but it is more difficult to make.

4 SCARFED JOINT This joint is the most efficient because there are no stress concentrations at the ends of the overlap. It can be used in production, however, only when thicker materials are used. Moreover, it is difficult and expensive to make.

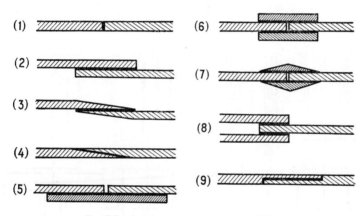

Fig. 7.2 Joints between sheet materials.

5 SINGLE STRAP JOINT It is used in practice most often when one of the surfaces must be smooth.

6 DOUBLE STRAP JOINT The strength is higher than that of joint (5). However, it is rarely used in production since neither surface is smooth.

7 TAPERED STRAP JOINT The strength is high because stress distribution is uniform. Its use is expedient only with extruded section straps—otherwise machining costs are too high.

8 DOUBLE (OR SUPPORTED) LAP JOINT This joint gives the highest shear values because there is no chance of peel at the end of the bond area. The most efficient utilization of materials is achieved when the thickness ratio is 1:2:1.

9 DOUBLE BUTT JOINT Machining costs are very high compared with the strength that may be obtained, since the strength of adherends is reduced by half. This type of joint is inadequate.

Fig. 7.3 Tongue and groove butt joints.

Tongue-and-groove butt joints are shown in Figure 7.3. They provide greater strength than flat butt joints, but they require machining and can only be used for bonding thick parts. An advantage of these joints is that they are self-aligning during assembly.

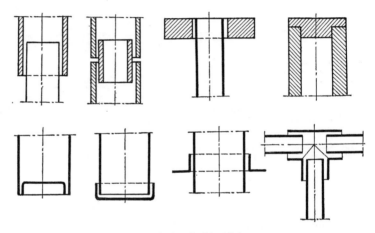

Fig. 7.4 Typical cylindrical joints.

Cylindrical joints, as seen in Figure 7.4, are most suitable for bonding parts such as tubing, shafts, and bushings, since when under tensile, compression or torsion loading, the adhesive layer is always subjected to shear stresses. In principle, only adhesives that do not require pressure for curing can be used for cylindrical joints. It is essential, that the parts are precisely aligned during bonding operations as the bond line thickness is determined by the difference in diameters. Some means for alignment are seen in Figure 7.5. Results from tests to elucidate the influence of adhesive thickness on the strength of cylindrical joints are shown in Figure 7.6 [2]. It is obvious that an increase in the clearance

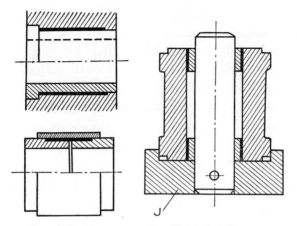

Fig. 7.5 Alignment of cylindrical joints: J – jig for alignment of bearing bushing.

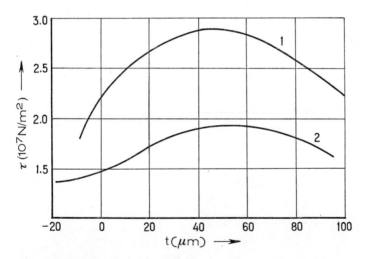

Fig. 7.6 Variation of shear strength (τ) of cylindrical joints with mean adhesive thickness (t) where t = one half of the clearance: 1 – for a heat-curing epoxide adhesive; 2 – for a room-temperature curing adhesive.

from -15 μm (interference) to 50 μm induces an increase in strength. The failure is of the adhesion type up to about $t=50$ μm, while it is of the cohesion type in the adhesive if $t>50$ μm. When $t>100$ μm, voids appear in the adhesive layer resulting from flowout during the initial stages of curing due to the decrease of capillary forces. The reduction in strength at clearances smaller than 50 μm could be explained by the greater influence of the irregularities of the adhesive thickness, the errors in the shape of adherend surfaces (for example, ellipticity), and inevitable excentricity. To avoid local stress concentrations the surface roughness height should be $h_{max} < (0\cdot 1 - 0\cdot 3)t$ (mm). The type and quantity of fillers in the adhesive should also be considered. It has been established that if opposite helical grooves with a trapezoidal cross-section (Figure 7.7) are cut into the cylindrical surfaces, the bond

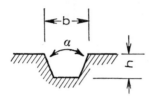

Fig. 7.7 Cross section through a groove in a cylindrical surface: $b = 0\cdot 5$ mm; $h = 0\cdot 3$ mm; $\alpha = 60°$.

strength increases by 10 to 15 per cent compared with smooth mating surfaces. However, the total surface of the grooves should not exceed 20 to 25 per cent of the bond area in order to avoid weakening the joint.

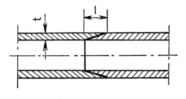

Fig. 7.8 Pipe joints with tapered ends.

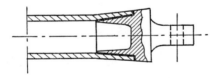

Pipe joints with tapered ends (Figure 7.8) have their advantages. Tapering allows the application of pressure during the cure and provides self-alignment of the parts. Taper depends on wall thickness; the best utilization of material strength is obtained if $l/t \geqslant 10$.

If a hole in one of the parts is blind, means for the escape of air should be provided for (Figure 7.9).

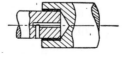

Fig. 7.9 Axle bonded in blind hole.

Large area joints are especially suitable for parts with greater thickness. But if one of the parts is thin, the joint is vulnerable to peel.

Angle and corner joints are subject to either peel or cleavage stresses, depending on the thickness of the metal parts. Typical designs are shown in Figure 7.10. Joint (1) may be used only in lightly loaded

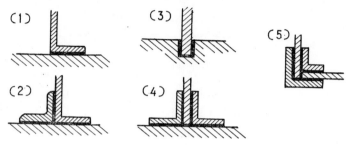

Fig. 7.10 Typical angle and corner joints.

structures. Higher strengths are attained by the use of supporting sections, (2), (4) and (5), or by fitting the end of one part into a groove (3). In corner joints between rigid parts adapted woodworking joint design (for example, end lap, mortise and tenon, mitre joint with spline) can be used. Figure 7.11 shows some types of bonds of corners of metal sections [3].

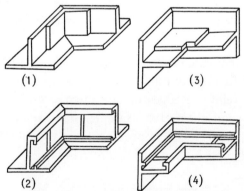

Fig. 7.11 Bonding the corners of metal sections. Joints (2) and (4) are secured against cleavage.

Bonding of stiffening members to large areas of thin metal sheets is used to reduce waviness, deflection and flutter. Some types of commonly used stiffening members, such as T-sections, hat sections and

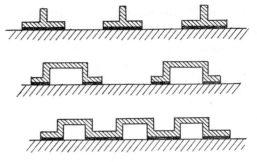

Fig. 7.12 Stiffener joints.

corrugated backing, are shown in Figure 7.12. To reduce stress concentrations, the ends of stiffening members should be made more flexible (Figure 7.13) and/or possibly secured with elastic (hollow) rivets.

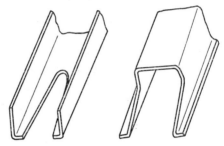

Fig. 7.13 Weakening the end of stiffening members to increase flexibility.

In adhesive bonding, efficient design is usually simplified because the designer has the freedom to reinforce structures wherever extra strength is required. To reinforce edges of holes and openings and to reduce stress levels around them, for example, bonded laminates may be arranged in different ways, improving the dynamic strength of the structure. An example is shown in Figure 7.14 [4].

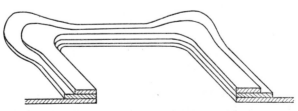

Fig. 7.14 Bonding laminates to reduce stress levels around a window.

7.2 Dimensioning adhesive bonded joints

Once a feasible bonded joint configuration has been chosen, the joint should be dimensioned. For the time being, however, generally valid formulae are still lacking owing to the variety of properties of the different adhesives and types of joints.

As far as simple lap joints between sheet metals are concerned summing up the empirical formulae for determining the distribution of stresses suggested by many authors [5], some of which are very complex (see Chapter 5.2), Müller [6] has shown that the mean bond strength can be expressed by:

$$\tau_{m_u} = B\left(1 + M \log \frac{\sqrt{t}}{l}\right)$$

where t is the sheet metal thickness, l the lap length and B and M are coefficients. This formula, however, is not convenient for practical calculations because the determination of the optimum lap length is very complicated. It is doubtful whether the data for the mean tensile shear strength will be of any use to the designer, since this strength is not a constant—it depends on the adherend thickness, lap length and other factors. The values for τ_{mu} may be used in design only in few special cases.

In the East German Central Welding Institute (ZIS) a simplified method for dimensioning simple lap joints on a purely empirical basis has been worked out [7]. They have drawn the conclusion that there is no point in using overlaps longer than those just able to initiate adherend yield, on the grounds that the joint will then, in any case, not be able to transmit higher loads. On this basis, an optimum lap length can be quoted for any given adhesive and adherend, rather than a joint failure stress at an arbitrary lap length. The method followed is explained below.

Test specimens with different lap lengths l are prepared and the tensile strength σ is determined. For a given adhesive the diagram $\sigma = f(l)$ is plotted (Figure 7.15) and the adherend yield stress σ_y (the 0·2 per cent offset yield strength $\sigma_{0·2}$) is drawn. The intersection of the σ_y line with the graph gives the lap length. It is clear that the adherend thickness and material are not changed. Plotting the results obtained in a diagram against σ_y ($\sigma_{0·2}$) for a given group of materials, e.g. light

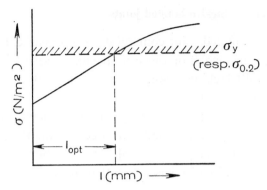

Fig. 7.15 Diagram for experimentally determining the optimum lap length for a given type of adhesive.

metals, for a given adhesive, e.g. a one-part heat-curing epoxide, and various adherend thicknesses, straight lines are obtained that can be expressed by:

$$l_{opt} = m\, \sigma_{0.2}$$

If the empirically determined coefficients m are plotted against the thickness t, in the case of the materials mentioned a parabola is obtained defined by the expression:

$$m = 0.2 t^2 + 0.2 = 0.2(t^2 + 1)$$

By substitution the following is obtained:

$$l_{opt} = 0.2\, \sigma_{0.2}\, (t^2 + 1)$$

To obtain l_{opt} in millimetres, $\sigma_{0.2}$ should be in 10^7 newtons per square metre and t in millimetres.

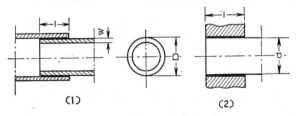

Fig. 7.16 Cylindrical joints.

For other adherend materials, adhesives and joint configurations, other equations are obtained, but the rules are always similar. For example, in the Table 7.1 formulae are given for dimensioning cylindrical joints as shown in Figure 7.16 [8]. The adherend material is light metal and the adhesive is a one-part heat-curing epoxide. For pipe joints the value t' is used instead of sheet thickness, defined by:

$$t' = w(D-w)/D$$

where D (mm) is the outer diameter of the smaller pipe and w (mm) the wall thickness.

TABLE 7.1 Formulae for calculating the optimum bond length in cylindrical joints

Type of joint	Load	Formulae
Pipe Fig. 7.16 (1)	Tensile	$l = 0.2\,\sigma_{0.2}\,(t'^2 + 1)$
	Torsion	$l = 0.2\,\tau_{0.2}\,(t'^2 + 1)$
Solid sections Fig. 7.16 (2)	Tensile	$l = 0.25\,\sigma_{0.2}\,(0.01d^2 + 1)$
	Torsion	$l = 0.25\,\tau_{0.2}\,(0.01d^2 + 1)$
	Compression	$l = 0.15\,\sigma_{0.2}\,(0.01d^2 + 1)$

Various safety factors must then be applied to the lap lengths obtained to allow for adhesive deterioration, fatigue, test errors, etc.

However, this apparently attractive method may, according to some authors [9], on closer inspection be not very reliable. No fail-safe behaviour can be available from joints designed on this basis, and the ability of an adhesive to hold a joint under conditions of destructive accident or damage is ignored.

From the above, it will be clear that despite the rapid growth of adhesive applications, since the basic knowledge of 'what makes things stick' is far from complete, designers must still proceed from design data received from adhesive suppliers, and rely on results obtained in their own investigations. But great care must be taken because often these data are not complete or meaningful and useful enough to be inserted directly into design calculations. For most adhesives, the standard tensile shear value is completely artificial and bears little or no relation to shear strengths obtained in large-area bonds and in most structures. Any comparison of tensile shear and lap data must, therefore, be made with the understanding that the tensile shear values are maximum values, and that in large-area joints the values are reduced. The percentage loss is influenced to a great extent by the elasticity of the adhesive. In practice, intuition will be indispensable for a long time yet in choosing the right adhesives and for determining joint geometries, safety factors and service lives.

In conclusion, Table 7.2 (published in [10]) is given, which contains the answers of experts, classified into three technological areas, to the question submitted by a journal to suppliers of industrial adhesives: 'Where do design engineers most often err in designing bonded joints?'.

TABLE 7.2 Where design engineers fail

Adhesive technology	Design consideration	Production line problems
Low peel strength	Using a butt joint when lap joint would be stronger	Lack of careful surface preparation
Overlooking such factors as pot life, curing time, operating temperatures	Loads causing unsuspected cleavage forces	Expecting prototype performance from bonds made on assembly line
Failure to get technical help from supplier in selecting an adhesive	Overlooking effect of increased service temperature in decreasing resistance to chemicals	Failure to keep surfaces clean until adhesive is applied
Assuming that strongest adhesive is always the best without considering cost or processing	Failure to check coefficients of expansion when unlike materials are bonded	Failure to consider the application method and equipment when designing joint
Lack of care in test procedures	Calling for heat-curing adhesive on a part that will not stand the heat	
	Overdesigning by asking for more strength or heat resistance than is needed	

7.3 Combining riveted, bolted and welded joints with adhesives for metals

One of the principal advantages of structural adhesives is their ability to minimize the fatigue problems associated with mechanical fasteners. Owing to this, adhesives can in many cases be used in conjunction with bolting, riveting or welding. This results in a considerable reduction or even elimination of stress concentrations.

Adhesive bonding is often combined with riveting if peeling or cleavage stresses are likely to develop in the joint. It is usually expedient to use hollow rivets, since their elasticity is higher than that of plain rivets.

Bonding combined with riveting results also in air- and liquid-tight joints and the elimination of capillary and contact corrosion. At the same time, improvement of fatigue life up to seven times without a commensurate increase in weight can be obtained. These factors contributed, for example, to the extensive use of the bonding-riveting technique for numerous fuselage areas and reinforcements in the Boeing 727 three-engine passenger jet; in this case a room-temperature-curing epoxide adhesive was utilized.

Different bonding–bolting systems have been tried in the field of structural steelwork [11]. Initially, bolts were only entrusted with the task of securing the bonded joint against peel and impact loads, and taking over the forces in cases of destructive accident (fire, for example). The first full-scale trial of this technique took place way back in 1955–56 when a pipe- and foot-bridge with a span of approximately 56 m was erected in West Germany. The tensile shear strengths determined during exploratory tests, using a polyester adhesive, were in the range of 1·1 to 1·6 × 10^7 N/m². Since under these conditions a connecting area 2·5 to 3 times larger compared with the areas used in conventional joining methods is needed, and since bolts have to be used as a precautionary measure, this technique was further developed. High-strength friction grip bolts were used instead of mild steel bolts, thus allowing prestressing of the joint, and the role of the adhesive was to increase the magnitude of friction between mating surfaces. This resulted in a significant increase in the load needed to cause slip, as compared with high-tensile bolt

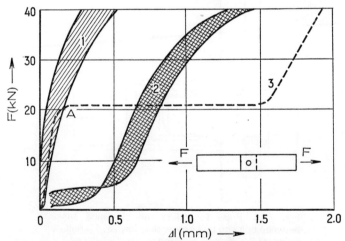

Fig. 7.17 Relationship between tensile load (F) and slip Δl) for steel specimens with bolt in simple shear: 1 – specimens incorporating filled epoxide-polyamide adhesive with mild steel bolts, 15·8 mm diameter tightened to a torque of 50 N.m; the 17·5 mm diameter holes are punched; 2 – specimens with turned and fitted mild steel bolts tightened to a torque of 50 N.m; the holes are drilled with a clearance of 0·25 to 0·45 mm; 3 – specimens with high-tensile bolts 15·8 mm diameter tightened to a torque of 210 N.m; the 17·5 mm holes are punched; A – start of slip.

joints without any adhesive, as well as a reduction of the amount of the slip. The efficiency may be improved by adding adequate fillers (corundum, for example) to the adhesive. Tensile shear strengths attained during static tests were in the range of 2·5 to 2·8 × 10⁷ N/m². Compared with high-tensile bolted joints without any adhesive, a 40 to 50 per cent proportion of bolt cross-section area is saved, as well as about 15 per cent of cross-section area in the case of connecting straps. Combining adhesives with prestressed high-strength bolting results likewise in markedly improving fatigue strength.

A further improvement in bolted joint efficiency consists in the use of an adhesive to fill the clearance space between bolt and bolt hole, thus minimizing bolt slip and relative movement between the joined members [12]. A higher loadability of the joint can be obtained than with turned and fitted bolts, due to the inevitable manufacturing tolerances associated with the latter. Tensile test results illustrated graphically in Figure 7.17 reveal also that whereas in specimens incorporating an adhesive, steel plate slip was still restricted at a tensile load of about 30 MN, in

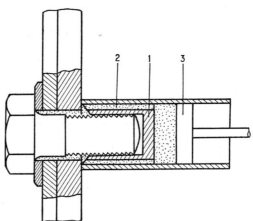

Fig. 7.18 Press for filling the clearance space between bolt and bolt hole with adhesive: 1 – hollow cylinder; 2 – slots; 3 – piston.

specimens with high-tensile bolts tightened at high torques, slip occurs when the load attains 20 MN, and continues until the clearance between bolt and steel plate is taken up. A simple press for filling the clearance space between bolt and bolt hole, as described in [13], is shown in Figure 7.18. The exchangeable hollow cylinder (1) is slipped over the screw

DESIGN OF ADHESIVE BONDED JOINTS 111

thread. There are axial holes or slots (2) along the periphery of the cylinder; the adhesive is pressed through them by the action of the piston (3) and enters the clearance space. The air escapes at the side of the screw head.

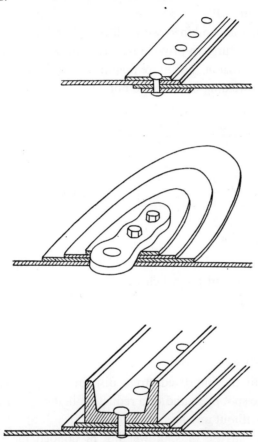

Fig. 7.19 Bonded doublers in riveted and bolted joints to reduce stress levels.

To reduce local stresses at mechanical joints in dynamically loaded structures the most commonly accepted method is either to increase the metal thickness locally by machining or chemical milling, or to attach doublers in the fastener areas. Increased weight, difficult sheet forming, and higher cost make the former method undesirable. With the latter method, riveting or spot welding doublers would neither minimize the local stresses nor give the desired appearance. The most practicable way to reduce local stresses without increasing sheet thickness is to bond certain details. Bonding is particularly advantageous for attaching very thin doublers which are adequate from the standpoint of stress distribution. Examples for reinforced fastener areas, as described in [4], are shown in Figure 7.19.

Spot welding is combined with adhesive bonding for joining in thin sheet metal structures to provide perfect sealing and eliminate capillary and contact corrosion. The more favourable performance, compared to joints by spot welding or adhesive bonding only, may be seen from the

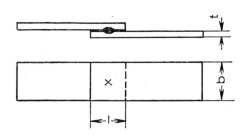

Fig. 7.20 Bonded and spot welded specimen. Nugget diameter = 5 mm, $t = 1$ mm, $b = 25$ mm, $l = 20$ mm.

results of testing bonded and spot welded specimens (Figure 7.20) under pulsating tensile load given below:

Type of joint	Fatigue strength ($\times 10^7$ N/m²)
Spot welded only	4·5
Bonded only	6·5
Combined	7·5

This is due to the improved stress distribution [14]. As a result, an increase in the spot pitch and/or a reduction in the weld nugget diameter are possible without any loss of strength [15].

The additional use of spot welding in bonded metal structures is recommended in areas where bending and peeling stresses may develop. Furthermore, fixing the adherends by spot welding can greatly reduce expenditure when bonding large structures.

Exploratory tests [16] have shown that CO_2 spot welding is in general not adequate; good results are obtained only by resistance spot welding combined with bonding.

The application of adhesive in the joint may take place either before or after welding. In the former case, there are two techniques. When using liquid or paste adhesives, they are applied on the pretreated sur-

DESIGN OF ADHESIVE BONDED JOINTS 113

faces of both parts to be joined, which are then assembled, and the welding is carried out through the adhesive layer [17]. Most of the existing structural adhesives may be used. However, it must be borne in mind that the filler type contained in the adhesive exerts a considerable influence on the quality of the welds [18].

When using tape adhesives, holes are cut out in them in the weld areas. The tape is stuck between the parts to be joined by heating, or wetting with a solvent. The welding is performed with templates having holes matching these in the adhesive tape; as a result, the welding contact area is free of adhesive.

When applying the adhesive after welding, liquid adhesives are employed. The parts to be joined are first cleaned and degreased, and then

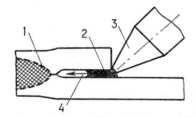

Fig. 7.21 Introduction of the adhesive into the spot welded joint clearance by the capillary method: 1 – weld nugget; 2 – adhesive layer; 3 – flow gun filled with adhesive; 4 – resultant force of capillary pressure.

welded. After that, the low-viscosity adhesive is applied to the edge of the welded joint using an applicator gun, as shown in Figure 7.21. The adhesive flows out under the action of its own weight. The penetration of the adhesive in the clearance between the parts is induced by capillary pressure [19].

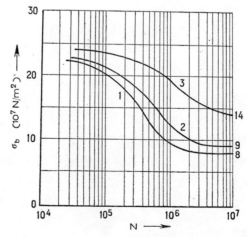

Fig. 7.22 Results of reversed bending fatigue tests for welded specimens: 1 – unmachined weld (with notches), $\sigma_e = 8 \times 10^7$ N/m²; 2 – machined weld (smooth surface), $\sigma_e = 9 \times 10^7$ N/m²; 3 – unmachined weld with epoxide adhesive coating, $\sigma_e = 14 \times 10^7$ N/m².

It is known that the fatigue strength of structures subjected to dynamic loading is considerably reduced by welded joints. This can be explained with the influence of the notches formed in the rough surface of the weld bead and the transition zone to the parent metal. An increase in strength may be achieved by machining the weld bead to obtain a smoother finish, but this would be difficult to realize in practical work and too expensive. But it has been found that, by applying a coating of an adhesive to fill up notches and irregularities in the weld, their effect can be minimized to a considerable extent [20]. Figure 7.22 shows fatigue test results for welded specimens without and with adhesive coating. It is seen that by applying an adhesive coating on the weld joint the endurance limit stress σ_e increases by merely 70 per cent under reversed stress.

There are three obvious possible explanations of this increase in strength (or service life), according to [21]. These are:

1 That the adhesive actually carries load and thereby 'softens' the stress concentration at the notches.
2 That the coating inhibits attack by the atmosphere.
3 That the bond between the adhesive and the metal is such as to delay crack propagation.

8
Application of Adhesive Bonding of Metals

8.1 Applications

In view of the tremendous number of adhesive formulations commercially available, and their spread and usage, it is impossible to cover all applications within the scope of this chapter. Consequently, only typical examples from different fields of application will be discussed. This collection should also give hints for new ways of designing and simpler fabrication.

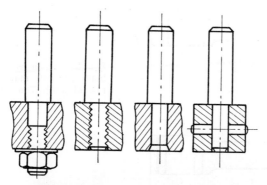

Fig. 8.1 Different mechanical means for attaching a stud: with a nut, by screwing on, by riveting, and with a pin.

A typical cylindrical joint, the attaching of a load-carrying stud, is illustrated in Figures 8.1 and 8.2. It is clearly seen that in this case bonding is the most simple joining method compared with the others. Failure in the bond occurs at a load $F=14$ kN if a room-temperature-curing epoxide adhesive is used, and at $F=25$ to 28 kN if a heat-curing epoxide adhesive is used.

Another example, often seen in practice, is the lever with bearing bush shown in Figure 8.3. Formerly, the bush was pressed and secured with a

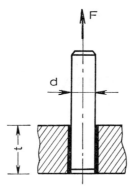

Fig. 8.2 Stud bonded to a plate ($d = 12$ mm diameter, $t = 25$ mm). The bond area is $A = 1,000$ mm².

screw; this required additional reaming. Now, the bush is bonded, a mandrel being used for alignment during the cure. The force needed to cause bond failure is about $F = 30$ kN for epoxide adhesives of the room-temperature-curing type, and $F = 70$ to 100 kN for the heat-curing type.

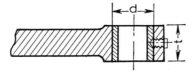

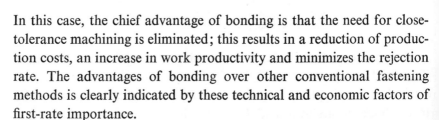

Fig. 8.3 Lever with bearing bush ($d = 30$ mm diameter, $t = 38$ mm). The bond area is 3,580 mm².

In this case, the chief advantage of bonding is that the need for close-tolerance machining is eliminated; this results in a reduction of production costs, an increase in work productivity and minimizes the rejection rate. The advantages of bonding over other conventional fastening methods is clearly indicated by these technical and economic factors of first-rate importance.

Figure 8.4 illustrates another example of cylindrical joints [1]. It deals with the housing of a valve for liquid fuels. In this case adhesive bonding also permits the use of a cheaper casting material. The whole housing was previously cast in bronze, since this material is most suitable

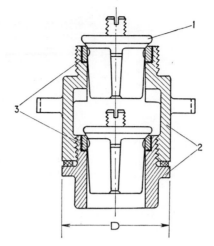

Fig. 8.4 Double valve for liquid fuels, $D = 54$ mm diameter: 1 – brass valve cone; 2 – housing; 3 – bonded brass seat rings.

to provide sealing at the seats. Whether it was justified, owing to this reason, to cast the part in bronze, had to be faced; as is known, better sealing is achieved when rolled brass is employed. Moreover, a high rejection rate (up to 80 per cent) resulted from casting the valve housing in bronze. The rejection rate was reduced to a minimum by using a composite design: a cast iron housing with bonded seats, made of rolled brass. Moreover, quantities of non-ferrous metals are saved.

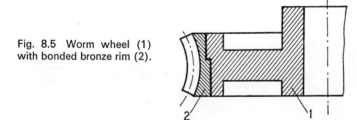

Fig. 8.5 Worm wheel (1) with bonded bronze rim (2).

The bronze rim of a worm wheel, as seen in Figure 8.5, is bonded to a steel body. This again is an example of a cylindrical joint, where important savings in bronze may result. Savings are made even if the rim has previously been pressed or shrunk on, since these techniques require a greater thickness of the rim to withstand the stresses induced. In bonding, there is no load acting on the rim. Since only a coarse fit is needed when bonding, the rim must be bonded to the wheel before the teeth are cut.

118 METAL-TO-METAL ADHESIVE BONDING

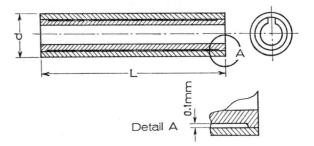

Fig. 8.6 Switch shaft with key-way: $d = 13$ mm, $L = 70$ mm.

Figure 8.6 gives an example for simplifying the make of a key-way in small bores. Normally, such key-ways are made on key-waying or key seating machines; but this is only efficient up to a certain inner diameter

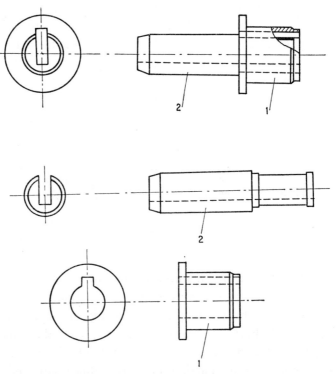

Fig. 8.7 Workpiece-holding mandrel and broach guide for an internal broaching machine. (*VEB Blechbearbeitungsmaschinenwerk, Aue/Sachs.*)

and bore length. Both brass tubes are bonded together with a room-temperature-curing epoxide adhesive after being machined.

Figure 8.7 shows a steel bush (1) bonded to a steel mandrel (2) with a room-temperature-curing epoxide adhesive. Depending on the diameter, up to 50 per cent savings in material are achieved. Moreover, the machining time for turning and key-waying is greatly reduced.

Figure 8.8 illustrates, in principle, an axle with a journal, this is a telling example of how bonding can result in savings in material and

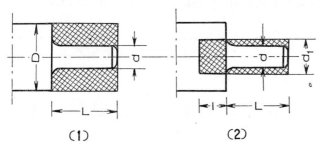

Fig. 8.8 Axle with journal: d = 25 mm.

machining. Data on the utilization of material and machining factors for several sizes in making it of solid material, as shown in Figure 8.8(a), are:

Diameter D (mm)	Utilization of material (%)	Machining factor (%)
40	39	61
50	25	75
60	17·5	82·5

In this case, the material utilization factor is the ratio between the weight of the finished and that of the semi-finished part; here it is $100 \, (d/D)^2$ %.

If the journal is made separately and is bonded to the axle, a 28 mm diameter steel bar can be used. In this case we have:

Diameters (mm)	Utilization of material (%)	Machining factor (%)
(a) For the axle		
$D = 40, d_1 = 28$	51	49
$D = 50, d_1 = 28$	68	32
$D = 60, d_1 = 28$	78	22

(b) For the journal:
$d = 25, d_1 = 28$ 82   18

The following data are valid in bonding:

Diameter of the bore: maximum 28·15 mm for $d_1 = 28$ mm
Bore depth: $l = 40$ mm
Bond area: $A = 3500$ mm²

If a room-temperature-curing epoxide adhesive is used, the failure force will be about 35 kN, corresponding to a shear strength of $\tau_u = 10^7$ N/m².

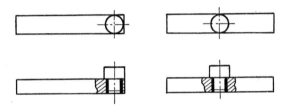

Fig. 8.9 Feather keys.

The feather keys shown in Figure 8.9 have previously been made from solid material, which required considerable machining. Adhesive bonding greatly reduced machining.

Steel tubes with outer diameters of 22 and 30 mm and a length of 2,500 mm are used in machines for spraying vineyards and orchards

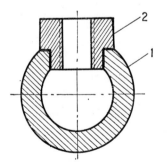

Fig. 8.10 Cross-section through a steel tube (1) with bonded nipple (2).

with pest and disease control preparations [2]. Only on one of the sides 5, 7, or 9, steel nipples with an internal thread for the spray nozzles are attached (Figure 8.10). Earlier, these nipples were welded, which offered much difficulty. Bonding of the nipples with an epoxide adhesive eliminates expensive and labour-consuming welding operations, subsequent straightening due to distortion caused by the high heat of welding, and greatly improved appearance.

Composite bearings are commonly made by lining a shell with special anti-friction metals, and, in certain cases, by pressure welding. However, these methods require special machining of the shells to obtain the fusion with the lining needed. The shells must be of comparatively pure low-carbon steel. Generally, composite bearings are produced efficiently only in specialised shops.

It has been established, however, that bonding (with epoxide adhesives, for example) will provide an adequate bond between the shell and a bearing bush [3]. Yet, since adhesives for metals usually possess a limited heat conductivity, its destruction is possible when heated. For this reason, to obtain better heat dissipation, a direct metal contact between the shell and the bush should be provided. This is effected by applying the adhesive in flat slots cut helicoidally in the shell or the bush. In this case, the bush can be made of drawn tubes and considerable savings in tin are achieved. Normal cast iron instead of high-grade steel may be used for the shell. The bushes are turned on a lathe, dry machining doing away with the need for subsequent treatment of the surfaces to be bonded.

A manufacturer fits cylinder liners in diesel engine blocks using a one-part anaerobic adhesive to reduce manufacturing costs and produce a stronger assembly. The adhesive is applied to hold the liners in position during final machining and to provide better retention and seal in service. No lip or countersink is required to locate the liner.

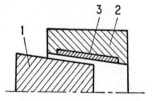

Fig. 8.11 Fitting a bevelled shaft (1) to a hub (2) using an adhesive (3).

In connecting machine parts, bevelled shafts in particular, the time needed to obtain a good fit can be greatly reduced, compared with common scraping, by using an adhesive (an epoxide for example) [4]. The adhesive is applied to the bore of the hub, and the greased shaft (with silicone oil, for example) is driven in (Figure 8.11). After the adhesive is cured the shaft is driven out to inspect the fit; then the parts are definitely assembled. The recess in the hub is provided in order to avoid the edge pressing on the adhesive.

Recently, a number of electrical and engineering works have begun to use bonding and embedding with synthetic resins for making cutting tools, mills, stamps, etc., since this greatly simplifies production. Mechanical clamping of ceramic inserts in tool-holders is not satisfactory in certain applications, boring small holes in particular. Because of its low coefficient of expansion, tungsten carbide can not be readily joined to metals of different thermal expansion rates at the elevated temperatures required by soldering techniques; this is due to the thermal strain and distortion introduced into the part during cooling.

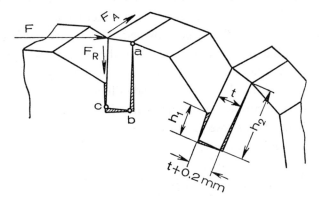

Fig. 8.12 Bonded mill cutting tools ($h_1/h_2 = \frac{1}{2}$ to $\frac{1}{4}$): F – cutting force; F_R – radial component of F; F_A = axial feed component of F.

Ceramic and metal cutting tools can be bonded to holders, mills, cutter heads, etc., by using adhesives such as epoxides which cure without heat, or in the relatively low temperature range of 100 to 200°C. Light sand-blasting and cleaning is needed before application of the adhesive. The bond strength should be great enough to bear the cutting forces. The cutting tips should be placed in such a way that the bond line is subjected only to shear or compression loads. The adhesive should be resistant to the temperatures developed during machining since the bond strength rapidly decreases at elevated temperatures. However, the heat conductivity of ceramic materials is low and the bond line cannot be heated excessively. Bonded cutting tips are easily removed, if necessary, by locally heating the assembly to above 130°C.

Figure 8.12 shows, as an example, bonded mill cutting tools [5]. Direct transmission of the cutting forces to the tool body takes place at the points (a), (b) and (c).

A manufacturer bonds the steel forged heads of light hammers to the glass fibre handle using a small amount of a one-part anaerobic adhesive. Tests have shown that such hammers withstand 300,000 blows successfully, thus exceeding by a factor of three the normally expected service life.

The unit shown in Figure 8.13 consists of a spring steel arm fastened by three hot-upset rivets to a zinc die casting, as described in [6]. It is

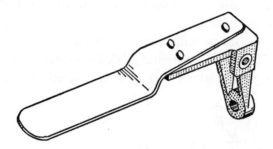

Fig. 8.13 Riveted assembly.
(*IBM, Rochester, Minn.*)

used in aligning data processing cards. After limited service, however, fatigue cracks developed in the casting between the rivets, due to the flexing of the casting in this area, necessitating replacement of the unit. To stabilize the casting and prevent flexing, a drop of a one-part epoxide adhesive is put between the detail parts before riveting the assembly. The rivets are then conventionally hot-upset, and later the adhesive is cured in an oven. No tooling is required for bonding, since the rivets serve as the necessary fixture. The results of bending fatigue tests and service experience have shown that the addition of the adhesive increases the life of the assembly by a factor of five to six, while requiring little change in manufacturing technique.

Bonding has also improved the working life of fans which, previously assembled by spot welding, used to develop fatigue cracks in service.

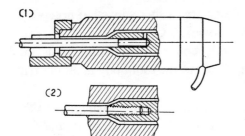

Fig. 8.14 Magnet with high switching frequency: (1) new bonded construction; (2) screwed construction.

Figure 8.14 shows a magnet with short stroke and a very fast switching rate, connected to the control slide valve of a hydraulic system. The rod, previously screwed on to the magnetic core, broke in the end of the thread after about 100,000 switching operations due to notch action. After bonding the rod in the magnetic core, it withstood more than 200 million switching operations.

Cast iron gate valves with bonded construction, as seen in Figure 8.15, are produced in W. Germany [7]. Compared with conventional design, a strong impression is left by their aesthetical appearance. This

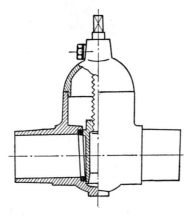

Fig. 8.15 Cast iron gate valve: diameter 80 mm, allowable pressure 10×10^5 N/m². (*Doering GmbH, Sinn, Dillkreis*)

results in better protection from corrosion, particularly in underground installation, since there are no recesses and edges. The number of parts is greatly reduced and machining is simplified. The weight of the new valve is about 45 per cent lower than that of similar conventional constructions. Bonding is carried out at a temperature of 100°C. If it is necessary to disassemble a valve for reconditioning or repair, it must be heated to a higher temperature at which the adhesive will loose strength, thus making disassembly easier.

APPLICATION OF ADHESIVE BONDING   125

Adhesive bonding permits complicated parts to be made easily and economically. For example, simple die castings can be bonded together after machining to form complicated assemblies. A large pump manufacturer has, by using this technique, reduced rejection rates to nearly zero from a previous high of 25 per cent in the manufacture of pump castings. As reported in [8], in one application two separate die castings are assembled by bonding to form a complicated pump assembly. The adhesive used is a one-part epoxide. Previously the pump part was cast in one piece by the sand-mould process. However, because of the complex interior design of the casting, blow-holes often occurred and resulted in rejection rates as high as 25 per cent. The use of an adhesive eliminates the need for lugs, bolt holes and flanges necessary with mechanical fastening, as well as separate sealing operations.

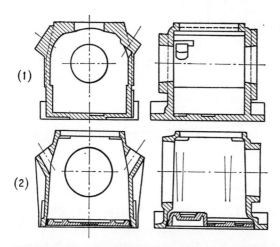

Fig. 8.16 Compressor housing: (1) gravity die cast; (2) pressure die cast. (*Westinghouse Bremsen GmbH, Hanover*)

Formerly, the housing of the one-cylinder air compressor shown in Figure 8.16 was gravity die cast from aluminium in one piece (1). The new construction (2) is cast under pressure and the bottom is subsequently bonded into position using a heat-curing epoxide adhesive [9]. In addition to reducing the rejection rate by about 45 per cent, significant savings resulted because of less machining needed.

The tube ends in tube-bundle heat exchangers can be bonded into the tube plates with epoxide adhesives instead of being soldered or radially rolled. This method does not limit the choice of material, simplifies production and provides a really solid seal. The tube ends are sandblasted and coated with the adhesive. With a predetermined clearance, they are then inserted into the drilled tube plate (the holes of which have also been coated with adhesive) and are moved about to work the adhesive down the holes. The method has been developed further by dispensing with a conventional metal tube plate or header, and casting one directly on to the bundle of tubes using epoxide resin [10].

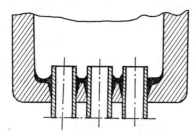

Fig. 8.17 Cross-section through an oil separator with bonded aluminium tubes.

A similar problem—the fastening and sealing of aluminium tube ends to cast aluminium headers of oil separators—has been solved, unsatisfactorily, by shielded arc welding, the rejection rate being very high due to porous welds. Owing to this, a compressor manufacturer proceeded to bond the thin-walled aluminium tubes to the headers (Figure 8.17). The results were excellent [11]. A powder epoxide adhesive was used. It was filled up at room temperature in the header and distributed around the tubes, which protrude about 5 mm into the header, by vibration. The holes for the tubes were bevelled to ease the penetration of the adhesive. The latter was then melted and heat-cured at 180°C. After curing, the adhesive had the same metallic appearance as the bonded aluminium parts. The tubes were exposed to hot oil vapours in service.

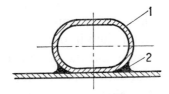

Fig. 8.18 Detail of evaporator cooling coil for freezers.

The manufacture of refrigeration equipment may also be simplified by bonding and production costs reduced. Figure 8.18 illustrates as an example a detail of an evaporator cooling coil for domestic refrigerators, as described in [12]. In the bonding, the aluminium tube (1) is pressed to an oval shape. At the same time, the adhesive (2) is laterally extruded, thus achieving a metal contact between tube and body. Measurements have shown that heat conduction was good.

It has been established that the Freon 12 and Freon 22 gases do not affect epoxide adhesive layers on metal surfaces. This gave the idea of bonding a number of parts of Freon compressors.

Another application in the refrigerator field is the bonding of aluminium and copper U-tube joints on heat exchangers. The conventional method of brazing or silver soldering each individual joint is time-consuming, whereas a number of adhesive bonded assemblies can be placed in an oven and cured in one operation.

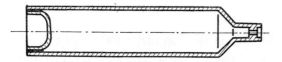

Fig. 8.19 Pressurized tube for butane fuel.

For pressurized receptacles for liquefied gas adapted for recharging gas lighters, according to [13], a tube is used, comprising a cylindrical body of cold-impact-extruded metal (Figure 8.19). This body has a substantially uniform bore and walls tapering from one end to the other, thereby providing a relatively thick wall at the open end. The open end is closed by an annular plug using a sealing composition. Pre-formed epoxide adhesive pellets may be introduced into the tube, and on heating, they seal and bond in one operation. The strength and resistance of the seal to fracture is greater than that of the thin wall at the opposite end of the body. If excessive pressures develop in the container the said thin portion of the wall will fail before the adhesive joint gives way.

Epoxide adhesive is also used for attaching steel filter screens covering the cooling slots in aluminium-cast housings of welding tractors (Figure 8.20).

There are many cases when bonding is used to temporarily join parts to make machining and/or assembly easier. For example, in lathe turning piston rings or other similar parts, a great number of rings can be bonded together using a soluble adhesive to form a cylinder. For better centring special centring rings are bonded to the faces of the cylinder [14]. In East Germany epoxide adhesives are used to bond the blades of hydraulic machines to special disks for easier machining [15]. One American engine manufacturer bonds a spacer and two piston rings to facilitate assembly; in service, the engine oil dissolves the adhesive and the parts separate and function as intended.

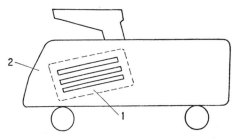

Fig. 8.20 Bonded filter screen (1) over the cooling slots in the aluminium-cast housing (2) of a welding tractor.

According to [16], in splicing metal foil strips (1), as seen in Figure 8.21, a tape is used comprising a layer of foil (2) (0·01 mm aluminium foil, for example) coated with two types of adhesives. The coating (3), disposed centrally, is tacky at room temperature and first secures the bond. The spaced bands (4) are of a second adhesive which is heat-responsive in the range of temperatures utilized for annealing the foil strip to form a strong bond between the tape and the foil strip being spliced, and to prevent the central adhesive bands from 'bleeding'.

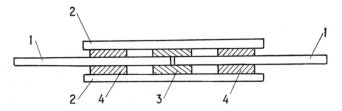

Fig. 8.21 Dual-adhesive splicing system (exaggerated) : 1 – metal foil strip being spliced; 2 – aluminium foil; 3 – room-temperature curing adhesive; 4 – heat-curing adhesive.

Designers now accept the method of bonding friction lining material to clutch plates and brake shoes in a variety of mechanical drive systems [17]. This method offers many advantages. Rivet holes in the linings and plates (shoes) are not required, and the linings are attached more easily and cheaply. Elimination of the rivet holes results in a larger friction, i.e. braking area on the linings, improving their efficiency. By dispensing with rivets, the danger of scored brake drums is eliminated. Heat dissipation is better, since the lining is in full contact with the plate (shoe) over its whole area. In addition, bonded linings can be safely used to a much reduced thickness compared with riveted linings, so that re-lining is less frequently necessary. The adhesives used for bonding friction linings are phenolic or epoxide based. They must provide a reliable bond throughout the temperature range of −65°C to over 350°C. Obviously, in addition to heat resistance, these adhesives must be extremely resistant to petrol and oil; for example, in the case of motor car transmission applications they are constantly bathed in hot oil.

Since the first bonding of tank clutches in 1944, many thousands of brake and clutch assemblies have been made. Among the items now being bonded daily are brake shoes for automobiles, lorries and buses, clutch plates for military vehicles, high-performance sports cars, tractors, earth-removing machinery and machine tools, brake bands for drag line equipment of up to 9 tons grab, brake bands for tractors and the brakes on trawl winches. Some of these components are of considerable size. For example, some drag line equipments employ brake bands of up to 1·5 m diameter.

Experience has shown that most re-lining of automobile brakes by bonding can be carried out more efficiently by service depots or larger brake re-lining specialists than by the smaller garage unit. The chief reason is that a certain amount of equipment is necessary and smaller shops might not find it justifiable to provide the necessary outlay for this kind of work.

Bonding of friction materials excepted, the use of adhesives in automobiles [18] is extremely conservative compared to their use in aircraft. One problem facing the automobile industry is the fluttering and rattling of engine bonnets and boot lids at high speeds or on rough roads. This is due to the insecure fastening, the reinforcements being welded to the covers and lids only along the outside edge. Some automobile manufacturers have solved this problem by the use of a heat-curing adhesive to bond the ribs of the reinforcement members to the underside of covers and lids. This method has not only eliminated flutter, but has increased design flexibility, reduced the weight of the assemblies, and has cut manufacturing costs. Semi- or fully automated machines are used to apply the adhesive at about 100 spots on the underside of covers and lids [19]. Reinforcement members are then laid and clamped in place (Figure 8.22). The adhesive cures to its maximum strength during a subsequent phosphatising and priming operation or during the paint baking. Although the surfaces bonded are not usually degreased, a shear strength of about $0 \cdot 5 \times 10^7$ N/m² is obtained.

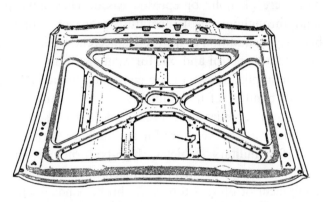

Fig. 8.22 Automobile engine bonnet. The points at which drops of adhesive are applied between the outer and inner panels are shown. (*Ford Motor Co.*)

Recently, some automobile manufacturers have used adhesives to bond roof bows to the roof to reduce noise and to increase rigidity. Epoxide–nitrile rubber adhesives are used to bond and seal glass to zinc die cast or stainless steel channels in vent and side windows of automobiles. They are applied as an extruded tape in the channel, into which the glass is inserted. Curing is accomplished with heat, while the assembly is held in a jig.

Other uses of adhesives in automobiles might be considered minor from a volume standpoint. For example, adhesive attachment of weather strip to door flanges provides necessary sealing. Adhesives make possible many of the decorative uses of metals and plastics in what otherwise would be incompatible combinations. Clear and coloured plastics are joined to steel, aluminium and chrome-plated parts with selected adhesives to give artful designs and insignia. Polyamide hot-melt adhesives are used for attaching trim on to stove-enamelled metal surfaces, thus eliminating mechanical fasteners.

In automobile applications, the sealing or gap-filling properties of an adhesive may be just as important as the joining function. Synthetic rubber-based expandable sealants are applied to joints between sheet metal parts that are spot-welded and subsequently become inaccessible for sealing. They expand at paint-bake oven temperature to form waterproof and rustproof joints.

External sliding doors for railway passenger carriages are fabricated in bonded construction in East Germany [20]. The door structure consists of several stiffening U-sections to both sides of which two facing sheets are bonded using a heat-curing phenolic adhesive (Figure 8.23).

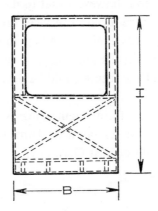

Fig. 8.23 Sliding door for railway passenger carriages: $B = 1{,}420$ mm, $H = 2{,}035$ mm.

By skipping the welded steel sheet construction, an AlMg-alloy bonded construction is employed, thus giving a significant reduction in weight, e.g. from 120 to 50 kg per door. As a result, the service life of door locks is increased considerably. In spite of higher material costs, as compared with steel, significant cost and production time savings result due to reduced and simplified preliminary work, no need of corrosion-protective paints, simplified door lock mounting, etc. Before regular production was started, several doors were tested under severe end-service conditions. A special lever system and pre-stressed rubber ropes opened and shut each of these doors on a special stand at a rate of 40 times per minute; after one million cycles the test was discontinued.

A 'wheel-axle set of the future' has been recently developed for the West German Federal Railways. The idea was to eliminate the well known disadvantages of conventional steel wheel-axle sets: high weight, high internal stresses caused by shrink and press joints, and bell-like ringing noises. In the design chosen the axle shaft is hollow and is built up of a number of thin-walled tubes placed and bonded into one another. The forged steel axle journals are bonded into the axle shaft. The wheel body, similarly bonded to the axle shaft, is made up of two symmetrical semi-shells of glass-fibre-reinforced epoxide resin. Adhesive is also used to bond the high-strength light metal rim to the wheel body. The new wheel set has a weight of 500 kg only and a load carrying capacity of 20,000 kg; the weight of a conventional wheel set of the same load carrying capacity would be 1,260 kg, thus saving 60 per cent in weight. Dynamic tests have shown the high reliability of the new set. It should be borne in mind, however, that in its present form the new wheel set still cannot be introduced in service. Provided progress is achieved in other fields of railway engineering (e.g. lower carriage weights achieved by the use of light metals and plastics, substitution of the block brakes), the wheel set of the future could be developed into a product for large-scale use.

Bonding in farm engineering is still mainly used in repair work. Figure 8.24 shows bonded elements of mowing machines successfully experimented in East Germany.

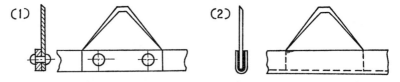

Fig. 8.24 Knives for mowing machines: (1) riveted construction; (2) new bonded construction.

In the electrical and electronics industry cylindrical bonded joints are used for the fixing of axles, sockets, ceramic shields for spark plugs, etc. Bonding replaces the pressing of collector rings on the shafts of small electric motors; this eliminates frequent cracking of collectors, and the need of a collector steel bush.

The attachment of silver-coated mirrors in oscilloscopes, rapid ciné cameras and other optical instruments by bonding is nothing new. Other examples of application are the bonding of permanent magnets, of magnetic materials to iron yokes in measuring instruments, of loudspeaker magnets, and of high-fidelity cartridges.

APPLICATION OF ADHESIVE BONDING 133

The bodies of magnetic chucks (for planer-type surface grinders, for example) can be formed of alternate strips of steel and manganese-bronze or aluminium bonded together with an epoxide adhesive. Assembly is carried out without screws and clamps, and the magnetic flow is improved. This design provides a working face that remains rigid under heavy loads, even on large chucks. The bond is unaffected by chemicals, oils and solvents.

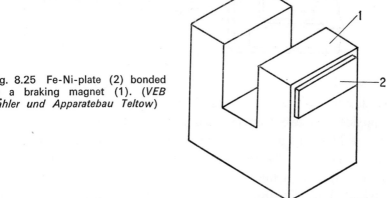

Fig. 8.25 Fe-Ni-plate (2) bonded to a braking magnet (1). (*VEB Zähler und Apparatebau Teltow*)

Figure 8.25 shows the Alnico braking magnet of a single-phase a.c. electrometer. The Fe-Ni-plate, providing temperature compensation during measurement reading, is bonded with a room-temperature-curing epoxide adhesive. The bond must withstand the baking temperature of the laquer coating on the magnet (about 150°C). In this case bonding was envisaged during design, because there was no other possible method of fastening.

Adhesive layers are normally good insulators. This property finds use, for instance, in the production of printed circuits by bonding foil or rolled copper to phenolic or other laminated plastics. The bond must withstand solder dip temperatures up to 250°C for 10 sec. One of the newer methods which lends itself particularly well to the mass production of printed circuits is that of die stamping [21]. Basically, this process utilizes a heated die which cuts a conductive pattern from adhesive-coated copper and simultaneously heat-reactivates the adhesive joining the metal to the insulating base-board material. This method makes it possible to use a far wider range of base-board materials than is commonly used for etched circuits because it involves no chemical solutions which might impair their properties. Furthermore, while the conductor metal used in etched circuits is usually limited to a maximum thickness of 0·1 mm, thicknesses up to four times greater can be used in the die stamping process, thus doubling the current-carrying capacity.

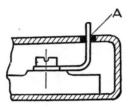

8.26 Insulated conductor outlet;
A – adhesive.

Conductor outlets, as seen in Figure 8.26, can be insulated with epoxide adhesives, which have a dielectric strength of about 20 kV/mm.

Epoxide adhesives can be used to create functional surfaces [22]. Figure 8.27 shows as an example a brass code cylinder with grooves milled 2 mm deep. These grooves are filled up with an adhesive containing anti-friction fillers. The surface is ground after the cure. The electrically insulated areas created by this method are resistant to wear by sliding contacts.

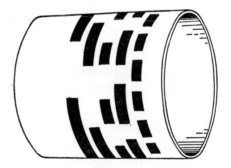

Fig. 8.27 Code cylinder.

The large-scale use of adhesives and resins for embedding coils, relays and other elements is worthy of notice. The advantages it offers are simplified design, protection against climatic conditions, insensitivity to impact and other external factors, and improved heat dissipation provided an adequate filler is added (quartz, for example).

The insulating properties of adhesives find application also in assembling and bonding laminated stacks for electric motors, transformers, and small assemblies (tape recorder heads and magnetostriction oscillators, for example). In practice, the following is done. Prior to stamping, a heat-curing adhesive is applied to both sides of each lamination with a spray gun or a simple roller applicator. Stamping is performed after drying at elevated temperatures to evaporate the solvents. The laminations are then placed in jigs and tightened down. The loaded jigs are placed in an oven to cure the adhesive. Uniform application of the adhesive on large surfaces is facilitated by the use of film adhesives.

Methods of joining metallic parts of electric motors and transformers in assembled relation are covered by the patents [23]. The methods comprise:

1. Positioning the metallic parts in fixed pre-assembled and overlapping relation to form a predetermined alignment and capillary spacing between the interfacial surfaces to be joined.
2. Applying a heat-curing adhesive (an epoxide, for example) at an edge of the capillary spacing of the interfacial surfaces.
3. Heating the assembly to a predetermined temperature to cause the resin to reach its gelation state. Heating the adhesive causes its viscosity to be lowered, with substantially all of the adhesive being drawn into the capillary spacing between the aligned interfacial surfaces to be joined by capillary action, to wet these surfaces before the gelation state is reached.
4. Curing the adhesive to a solid infusible state, thereby joining the metallic parts together.

The advantages of the method described are that it greatly reduces assembly time. A decrease in iron losses or a reduction of active material are achieved; thus the transformers obtained are more compact and cheaper. Noise is cut by 5 to 10 per cent. Scattering is reduced, which in turn leads to a decrease in current losses compared with iron bodies fastened by stud bolts or rivets.

Following a study of the busbar systems in general use, during which the requirements for efficient power distribution in switchgear cubicles were completely reappraised, a radically new system based on broad, light gauge, sheet copper conductors has been developed. This is claimed [24] to be a major advance in busbar design, giving up to 70 per cent saving in copper. The conductors, one per phase, are sandwiched between layers of insulant using an epoxide adhesive (Figure 8.28). The whole assembly is mounted in a jig and heat-cured under pressure. The fully sealed and insulated distribution system has many advantages. For example, it is self-supporting, takes up as little as half the space normally required, and needs virtually no maintenance. A complete busbar unit, although only 15 mm thick, will carry 3,000 A.

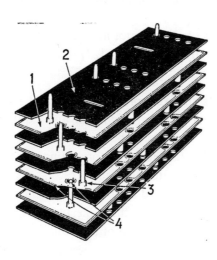

Fig. 8.28 Exploded view of the laminated busbar system: 1 – light gauge copper; 2 – insulating sheet; 3 – brazed stem connection; 4 – insulating spacer and edging. (*George Ellison Limited*)

Structural adhesive assembly has made possible the manufacture of the largest magnetic coils ever built for a 12·5 BeV Zero Gradient High Intensity Proton Synchroton constructed in the USA [25]. The synchrotron—a high energy accelerator in which the internal structure of the atom may be studied—consists of eight curved magnet sections made up of an iron yoke and a copper coil. The resultant ring magnet made by these sections is approximately 60 m in diameter. Each coil consists of 32 rectangular, water-cooled copper conductors insulated by and bonded to epoxide–glass-fibre laminates with a one-part epoxide adhesive. Adhesive assembly of the coils was the only practical way of achieving close conductor-to-conductor dimensional tolerances. In addition, the bonded assembly also makes a rigid monolithic structure which can be simply anchored in place.

In the manufacture of chokes for fluorescent lamps, adhesives are used for fixing the cores and filling the gap between them (Figure 8.29). Noise is considerably reduced since the rumbling, occurring previously in the unfilled air gap, is eliminated. Furthermore, the fixed size of the gap cannot change subsequently. A heat-curing epoxide adhesive is applied to both ends of the cores, which are then assembled; the distance

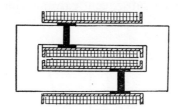

Fig. 8.29 Choke for fluorescent lamps.

between them is fixed with two screws according to the readings of a measuring instrument. The excess adhesive fastens the coils to the core.

Conductive adhesives are being used in many applications for joining metals where solder will not adhere, or heat of soldering cannot be tolerated. A number of conductive adhesives are available. Usually they are filled with different metal powders. Good results are obtained with nickel powder of 5 μm grain size, for example. Bond strength greatly depends on filler content and the optimum is achieved when the latter is about 30 per cent. Current load should not exceed 0·5 A/mm^2 and varies depending on cooling.

Adhesives are widely used instead of screws and rivets for attaching nameplates, instruction plates, and other types of indicator and data labels for machinery and consumer products. Of particular interest are the self-adhesive labels made of aluminium foil. Their surface is anodised and may be differently dyed. Individual data, such as serial and type numbers, dates, data for power, capacity, revolutions, voltages, etc., can be typed on a normal typewriter. The back side of the labels is coated with an adhesive, protected by a Cellophane or plastics foil; the latter is removed immediately before bonding.

Adhesive bonding is applied to the fabrication of electronic cabinets, chassis, and other enclosures. Compared to spot welding or riveting, bonding provides closer dimensional control without rework, improved appearance, simplified construction, gas tightness, and added strength.

If earthing is required, as well as to prevent peeling, additional fastening with a few rivets may be needed.

Figure 8.30 shows a fastening screw M 4 × 15 mm bonded to a front panel (steel or aluminium) for electronic instruments. A highly filled room-temperature-curing epoxide adhesive is used. When tightening a nut the thread will fail before the screw is loosened. The bond must withstand a subsequent galvanic zincing needed if the panel is of steel.

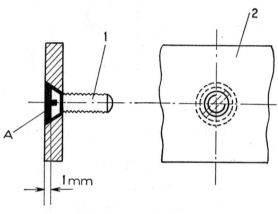

Fig. 8.30 Countersunk screw (1) bonded in panel (2); A – adhesive.

Previously, the countersunk screws were welded. Bonding, however, eliminates distortions by heat resulting from welding and the required smoothening down.

Glasses of measuring instruments are usually fastened to the enclosure using putty or laquer, rubber strips and springs, rings, eccenters, screws, etc. However, all these common methods are complicated and do not provide absolute water- and gas-tightness. Owing to this, fastening by bonding is expedient. The adhesive must have a good adhesion both to the glass and the enclosure material (metal or thermosetting plastics), and furthermore the flexibility to take up differences in coefficients of expansion. The glasses should not be fastened as shown in Figure 8.31(1), but according to Figure 8.31(2). The latter is better from the point of view of design since the adhesive layer is loaded in shear; furthermore, different elongation is less marked.

Implosion-proof TV image tubes are made by bonding a light-weight steel rim band around the edges of the tube face, thus eliminating the need for a glass safety plate in front of the tube; a heat-curing phenolic adhesive is used.

The recently introduced new one-piece car batteries feature a leak-proof case bonded with epoxide adhesive.

Adhesives find many uses in photographic equipment manufacture. In a typical camera there may be 30 different locations where adhesives are employed. These include lenses and mounts, nameplates, range-

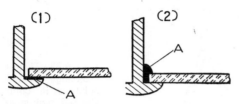

Fig. 8.31 Fastening a glass with epoxide adhesive (A).

finder mirrors, counting dials, covering, decorative strips, flash connection sockets and screw sealers.

It is more advantageous to make reels for ciné film and magnetic tape transport by bonding two flanges to a previously lathe-turned tube instead of turning the reels from bar material.

The packaging industry too is turning more to structural adhesives. Light-weight metal shipping containers are completely assembled by adhesive bonding, except for the fasteners. Adhesives achieve the strength desired, and also efficiently seal the containers without the use of additional sealing compounds.

The development of bonding techniques in the field of structural steelwork started as early as 1955–56, and soon reached the stage of industrial applicability. The first large fully bonded steel structure was a pipe- and footbridge with a span of approximately 56 m, erected over a canal in West Germany. Several smaller footbridges made from aluminium aloy sections bonded together with epoxide adhesives have since been built in Czechoslovakia and Austria; bolts had proved unsuitable, as deformation of the bolt holes in the light alloy soon appeared.

The use of adhesives in connection with pre-stressed high-strength friction-grip bolts for structural steel work provides a method of joining which is superior to bonding alone. This method was first tried on a large scale in 1963–64, when a three-chord pipe- and footbridge of 58 m span and 6 m over-all height with members inclined alternately in opposite direction (Figure 8.32), designed for a load of 90 kN/m, was erected in West Germany [26]. In the next couple of years 15 more bridges of similar construction with a span of 20 m were built.

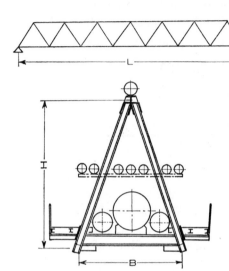

Fig. 8.32 Three-chord pipe- and footbridge assembled by the use of adhesive in connection with prestressed high-strength bolting: $L = 58$ m, $B = 4.4$ m, $H = 6$ m.

The method of improving bolted joint efficiency by filling the clearance space between bolt and bolt hole with an adhesive was first applied in Britain to a bridge with a span of 29 m erected across a canal at Uxbridge [27]. The bridge structure was made up from galvanised steel members bolted together utilizing punched bolt holes having a clearance of 1 mm on diameter. It was assembled with galvanized bolts using normal spanners, all mating surfaces being previously coated with a cold-setting epoxide-polyamide adhesive to increase friction at the interface of the members.

Combining adhesives with prestressed high-strength bolts as used for frictional connections in structural steelwork is also adapted for constructions of pre-fabricated members (composite constructions) [28]. Concrete slabs are bonded into position on to steel girders and are positively pressed by means of high-tensile bolts so as to increase friction between the mating surfaces (Figure 8.33). In order to obtain the shearing strength no special shear connectors are required (dowel-less connection). Instead of continuous cast-in-place concrete slabs, single precast prestressed slabs or planks can be put on. Using this technique, different composite constructions, as for example bridges, have already been erected in Britain, West Germany and Austria.

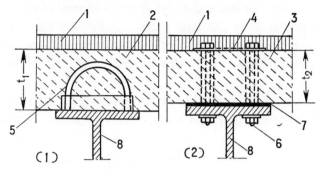

Fig. 8.33 Comparison between dowel construction (1) and dowel-less construction (2) : 1 – asphalt covering; 2 – cast-in-place concrete (t_1 = 200 to 240 mm); 3 – precast concrete slab (t_2 = 180 to 200 mm); 4 – pressure distributing steel plate; 5 – welded shear dowel; 6 – high-tensile bolts; 7 – adhesive layer; 8 – steel girder.

In composite metal structural members the favourable properties of two materials are combined [29]. Two materials very suitable for combination are aluminium alloys and steel, especially when a low weight and at the same time a small structural height and restricted deflection are required. These materials may also be combined in light-

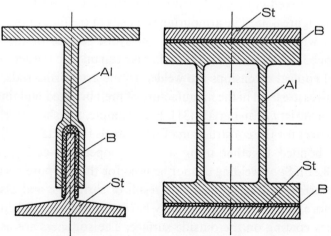

Fig. 8.34 Cross-section of bonded composite beams: Al – aluminium; St – steel; B – bond line.

weight structures subjected to rolling loads, since aluminium alone is not suitable. Another point of view would be, for example, the combination of the high corrosion resistance of aluminium with the low notch sensitivity of steel. In special cases, the high thermal expansion of aluminium can be compensated by the steel section. Adhesives may be used to provide a corrosion-proof bond of sufficient strength between steel and aluminium. Figure 8.34 shows two examples of composite sections.

Adhesives have been applied in several reinforced concrete constructions. As an example, a special method for butt joining load-carrying

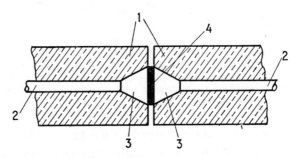

Fig. 8.35 Bonded concrete-armouring rods: 1 – concrete structural members; 2 – armouring rods; 3 – conical extensions; 4 – adhesive layer.

precast reinforced concrete structural members, as described in [30], is shown in Figure 8.35. The armouring rod ends in the face of the concrete members with a conical extension. The front surface of the latter is dimensioned in such a way that the adhesive can take up the acting forces. The steel conical extensions are welded to the armouring rods.

Adhesives are used in the manufacture of prefabricated building panels made of a variety of materials [31]. For example, the sheet steel facings and stiffeners for the construction of wall panels for curtain-wall facades may be bonded together using a room-temperature-curing adhesive (Figure 8.36). Spot welding cannot be used for this purpose because the smooth surface will be spoiled as a result of shrinkage and distortion; furthermore, the facing sheets usually have a previously applied enamel or plastics coating on the outside surface. The stiffeners are usually of galvanized steel. The core material can be foam plastics, mineral wool, vermiculite, etc., depending on the requirements. The inner facing may

APPLICATION OF ADHESIVE BONDING 143

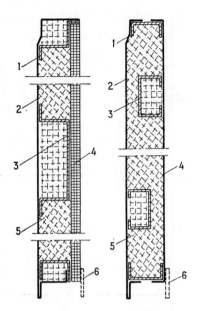

Fig. 8.36 Plastics-coated sheet steel wall panels with bonded stiffeners: 1 – frame section; 2 – outside facing; 3 – stiffener; 4 – inner sheet; 5 – insulation; 6 – clamping section.

be of pressure-moulded wood particle board, for example. In special cases, however, as well as in buildings of greater height, both facings of facade panels should be of sheet steel in accordance with regulations for fire resistance.

By bonding a metal facing to one or both sides of laminated plywood, substantial increases in flexural rigidity are gained, as shown graphically in Figure 8.37a; the curves in Figure 8.37b compare costs of such laminates with those of aluminium and steel plate, as illustrated in [32]. Virtually any combination of metal and plywood can be produced. Although laminates are usually used as flat sheets, panels with metal facing on one side only can be formed to a certain degree. Light-weight rigidity, as well as smooth, mar-resistant surfaces, make metal-faced plywood an excellent material for doors, railway carriage interiors, foundry pattern boards, and many others.

The use of adhesive bonding to assemble architectural aluminium and stainless steel window frames and sashes can greatly increase production capacity and considerably reduce manufacturing costs [33]. All corners of the window frames and sashes are mitred and joined by the use of die castings (of zinc, for example) bonded with an epoxide adhesive. This method utilizes a large bonding area and provides a very rigid and high-

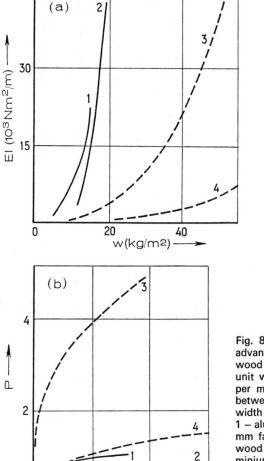

Fig. 8.37 Diagrams showing the advantages of metal-faced plywood: (a) relationship between unit weight (w) and rigidity (EI) per metre width; (b) relationship between rigidity (EI) per metre width and relative price index (P). 1 – aluminium-faced plywood (0·4 mm facings); 2 – steel-faced plywood (0·6 mm facings); 3 – aluminium; 4 – steel.

strength corner assembly. In addition, the adhesive seals the joints against moisture penetration.

Among the many types of double-glazed windows now available, the sealed units have proved themselves efficient and convenient. Installation is simple and, because there are only two sides exposed, cleaning costs are the same as for single-glazed windows. Metal-edge double-

APPLICATION OF ADHESIVE BONDING   145

glazing units can consist of two panes of glass separated by a metal spacer, e.g. a tube of square section. The glass is bonded to the spacer with an epoxide adhesive forming a rigid structure (Figure 8.38). The edges of the units can be protected by means of a mastic material (e.g. a butyl rubber composition), which is turn can be covered with a butyl-rubber-coated metal strip.

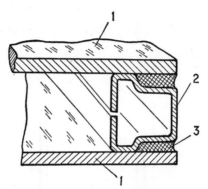

Fig. 8.38 Double-glazed window: 1 – glass; 2 – metal spacer; 3 – adhesive.

Bonding of steel pipes in sanitary installations in buildings instead of using threaded fittings to connect them permits the use of thin-walled pipes, since no increase in wall thickness to make thread-cutting possible will be needed. This may result in considerable material savings and in eliminating the labour-consuming welding of small-diameter pipes. Exploratory tests have shown that the best results are obtained when the two ends of the pipes are machined with an internal and an external cone, respectively, with a 1:16 taper.

Adhesive bonding provides lamp-posts of great strength. An American design of aluminium lamp-post utilizes a shaft made of two C-shaped solid extrusions joined together by means of an epoxide adhesive [34]. This type proved to be cheaper than similar structures made of seamless hollow extrusions which are in wide use.

A British firm fabricates the upper part of lamp-posts from three

Fig. 8.39 Cross-section through a lamp-post. (*AEI Lamp and Lighting Company Limited.*)

identical metal sections which are bonded together with an epoxide adhesive (Figure 8.39); they are then bonded to the bases, again with an epoxide adhesvie.

It is recommended in [35] that the light-metal parts of lamp-posts and flag-poles be fastened to other metal parts by bonding instead of welding, riveting or screwing; in this way neither mechanical strength nor appearance is impaired. Figure 8.40 shows, as an example, the fastening of an aluminium alloy pole shaft (1) to a steel base. The steel

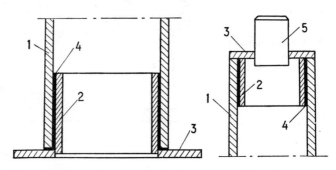

Fig. 8.40 Fastening the parts of lamp-posts by bonding.

ring (2) is first welded inside and outside to the steel disc (3); after smoothing down the weld, the base is cadmium-coated to eliminate corrosion, and the parts are assembled using a heat-curing adhesive (4). The fastening of the top cover with a stud (5) for the attachment of a lighting fitting is done in the same manner.

The laying of rails for underground railways immediately on the smooth floor of the tunnel by means of an adhesive, without using ties and gravel bed has proved reliable [36]. This construction (Figure 8.41) results in considerable savings, since the necessary height of the tunnel is 30 to 35 cm lower than usual.

A new fastening system has been developed in the USA for bolting the base supports of metal structures to concrete floors, especially in cases when sufficiently high loads cannot be achieved with conventional anchors. This new anchorage system uses a capsule containing a viscous liquid resin, a curing agent, and stone aggregate of controlled size. The capsule is inserted in a hole drilled in the concrete floor. Next a continuously threaded steel rod is placed into the hole which pierces the capsule and, by immediate rotation, mixes the contents to permit curing of the resin and subsequent tensioning of the rod. Curing time depends on the temperature and may be 30 min at 25°C.

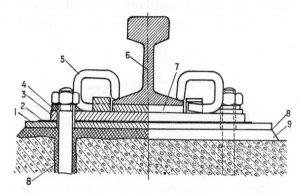

Fig. 8.41 Tie-less track in the tunnel of an underground railway: 1 – pressure distributing plate; 2 – insulating PVC layer; 3 – ribbed plate; 4 – insulating sleeve; 5 – clamp; 6 – rail; 7 – rubber interlayer; 8 – epoxide adhesive; 9 – epoxide–quartz sand mortar.

Epoxide resins provided a neat solution to the problem of bonding steel to rock encountered during the work in 1964–66 for saving the two colossal ancient Egyptian temples at Abu Simbel from the flood water which would be created by the new Aswan High Dam [37]. Both temples were cut bodily out of the sandstone rock, sawn into more than 3,000 blocks of not more than 30 tons weight, and then transported to a site 65 m higher up, above the future level of the water. There the whole structure has been carefully reassembled. Epoxide resin was used to anchor steel rods in the blocks so that they could be lifted for transport. Holes of about 35 to 45 mm diameter were bored in the rock and an epoxide resin mortar, containing sand as a filler, was poured into them. Knurled steel rods were then pushed right down into the holes and were firmly held when the epoxide resin set solid. The number and diameter of the rods used were varied according to the weight of each block. For blocks up to 20 tons, two rods were enough, but at least three were needed for blocks of 20 to 30 tons. After the curing of the resin, the blocks were ready for a crane to hoist them on to lorries for transport to the new site. After the re-erection work the ends of the rods were cut off.

All-metal skis are no novelty. Such constructions consist of two or three metal shells joined together by riveting and bonding. Figure 8.42 shows as an example a cross-section of the 'Aluflex' French skis made of high-tensile aluminium alloy. A heat-curing phenolic adhesive is used for bonding the shells together. Wooden inserts are provided in the area of

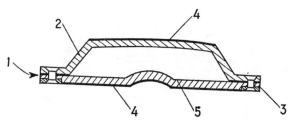

Fig. 8.42 Cross-section of an all-metal ski 'Aluflex': 1 – adhesive; 2 – aluminium, 2 mm; 3 – steel edge; 4 – vinyl coating; 5 – aluminium, 1·3 mm.

the bindings. However, only the introduction of composite constructions, combining the best properties of high-tensile alloys, wood and plastics (the latter as adhesive, laquer and running surface), handsomely contributed towards a metal ski breakthrough. The cutaway in Figure 8.43 shows a ski of this type having a sophisticated and complex design,

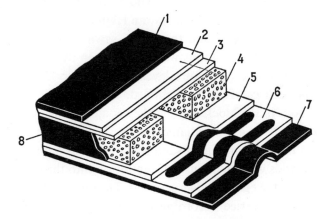

Fig. 8.43 Seven-layer bonded ski: 1 – phenolic fibre top sheet; 2 – high-carbon spring steel camber stabilizer; 3 – high-tensile aluminium; 4 – wood-particle board dampening core; 5 – high-tensile aluminium; 6 – perforated high-carbon spring steel sole, running the full length and width of the ski; 7 – textured phenolic fibre running surface; 8 – phenolic fibre side panels.

as described in [38]. The seven-layer bonded sandwich provides the ski with permanent camber, edges that never come off, and dampened spring action for better control at high speeds. This is achieved by bonding, by means of a modified phenolic adhesive, high-carbon spring steel under high temperature and pressure to aluminium, pressure-moulded wood particle board and plastics. The ski gains its over-all strength through the stressing of the seven layers during the bonding process. Aluminium thickness varies to produce skis with a different grade of stiffness—soft, medium and hard.

In view of the existing prejudices, it is strange that the use of adhesives for metal-to-metal bonding was first established in the aircraft industry, where requirements imposed on strength and security are very severe. In Britain bonding was accepted much earlier than in the United States. In 1943–44, De Havilland pioneered bonded airframe structures with the twin-engined Hornet fighter. Compared with conventional riveted constructions, adhesive bonding provides increased structural efficiency, smoother surfaces, better fatigue resistance, and the production advantages of fewer parts and fewer fabrication operations. In addition, significant weight and cost savings generally result. The weight savings are of great importance, since, owing to the increase in engine power and correspondingly in fuel consumption, the weight of aircraft is continuously rising. Surface smoothness contributes to higher speeds.

Because the bonded joint strength is so reliable, adhesives have now been employed to a greater or lesser extent in a large number of different aircraft, ranging from large passenger aircraft to guided missiles. In many of them bonding has served as basis in designing the complete airframe, joining metal to metal in primary structures such as wings and fuselage. Adhesives are used extensively to bond leading and trailing edges for helicopter rotor blades, integral wing fuel tanks, brake and clutch linings, and other assemblies. Much wider utilization of bonded structures is expected in the future, as better adhesives and quality control methods are developed.

As an example, over 55 per cent of the total surface area of the twin-turboprop airliner Fokker F27 Friendship is bonded, in over 400 assemblies with sizes ranging from 90 by 350 mm to tapered spar booms 10,000 mm long, and panels 1,300 by 4,900 mm. After extensive fatigue tests, including 14·5 million cycles of reverse loading on the wing assembly (equivalent to about 32,000 flying hours), unlimited life for the airframe under the fail-safe concept was granted. In the new twin-turbojet passenger aircraft F28 Fellowship, delivered to the airlines since 1968, as much as 75 per cent of the structure consists of bonded members.

Adhesives and bonded structures are also extensively used in the construction of satellites, where savings in weight are of paramount importance. However, for use in outer space special requirements are posed to the adhesives: they should not be influenced by ultraviolet light or gamma rays, and no degassing or other chemical changes that could lead to variations of the module of elasticity and bond strength should take place under high vacuum. For example, in each Tiros satellite, power for the equipment is derived from approximately 9,300 solar cells. They are mounted in 'shingles' of five cells on module boards carrying 80 cells each; the module boards, in turn, are attached to aluminium plates on the periphery of the satellite by using glass-fibre-supported epoxide adhesives [39].

Air-cushion vehicles must be light in weight in relation to their strength, and for this reason they incorporate structural forms which have long been used by the aircraft industry for conventional aircraft. These include bonded and metal honeycomb structures [40]. In the hovercraft built by Westland Aircraft Limited, constructed primarily of high-strength aluminium clad alloy, doublers and stiffeners are bonded to shear webs for support beams in the buoyancy tanks and for frame members in the deck and the superstructure. Buoyancy tanks, which form the basic load-carrying structure of the whole craft, are virtually all bonded to give a strong light unit which is completely watertight. These tanks have also to keep the craft afloat in the event of mechanical failure or when manoeuvring within the confines of harbours but must be light in weight in order that an economic payload can be carried. The lift fans are also bonded; each blade is fabricated from aluminium alloy sheet with bonded stiffeners giving a light component with good aerodynamic efficiency.

The use of adhesives for fastening metal parts of boats and barges is still limited. The joints between the hull members of small boats in particular are subjected to very light loads, and therefore, a wider application of adhesive bonding in boat building should be expected. It is important that adhesives used for this purpose be gap-filling to simplify the fabrication processes. Epoxide adhesives are being used, for example, to bond stiffeners to decks, transoms and skins of aluminium boats. If the section stiffeners are welded, the high heat of welding

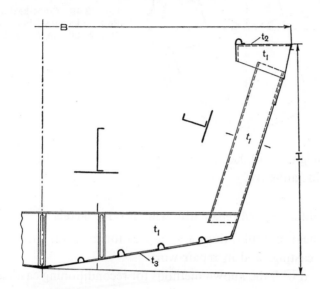

Fig. 8.44 Cross-section of a bonded motor boat hull: $B - 1{,}800$ mm, $H - 800$ mm, $t_1 = 2$ mm, $t_2 - 2 \cdot 5$ mm, $t_3 - 4$ mm.

causes considerable distortion of the structural members which require expensive straightening operations; adhesives eliminate this problem. Usually epoxides do not require pressure; but since the decks and skins are large and flexible, the parts should be clamped together to provide the contact pressure necessary to maintain an adhesive layer of about 0·15 mm thickness. Curing can be accomplished at room temperature, or faster if heat lamps are used. Figure 8.44 shows a cross-section of the hull of an aluminium motor boat built in the USSR [41]. Adhesive bonding has also been combined with spot welding to provide perfect sealing of the joints and to eliminate capillary corrosion.

Prof. Piccard's bathyscaph, shown in Figure 8.45, is another particular case. A record depth of 11,500 m was reached with it in the Pacific Ocean. It consists of three forged parts made of CrNiMo-steel, with an

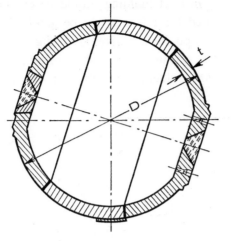

Fig. 8.45 Cross-section of Prof. Piccard's bathyscaph: $D = 2{,}180$ mm, $t = 120$ mm.

over-all weight of 12,000 kg, bonded together with a heat-cured two-part epoxide adhesive.

8.2 Reconditioning and repair with adhesives for metals

A special field of application of adhesives for metals is their use in the sealing of castings and in repair work.

Compared with the known methods for reconditioning castings which are of limited use, the method utilizing epoxide or polyester adhesives has advantages as listed below:

1. Internal stresses and structural changes in the parts, as caused by the heat of welding, are avoided.
2. Galvanic corrosion, likely to occur after soldering, is eliminated.
3. A better bond is provided between the part and the adhesive material as compared with the commonly used patching cements, putties, etc.
4. Expenditure on material and tooling is comparatively low; production time is also reduced.
5. Adhesives can be used in cases where welding torches, metal spraying guns and other tools cannot be used.

The difference should be borne in mind between filling-up of blow holes and large voids, that usually show after machining the surfaces and the impregnation (sealing) of porous castings where the pores are not visible and usually are revealed only after pressure tests.

In the first case, the voids are most frequently under the surface and do not affect the function and strength of the part. Firstly, the shape and size of the flaw must be established; for this purpose it may be necessary to enlarge the opening of the void—this will facilitate the filling-up with adhesive. The defective area must be cleaned and degreased; the adhesive is then applied. The part should be fixed in such a position as to help the separation of air. It is desirable to warm the part up to 40–60°C before applying the adhesive (this could be done locally with infra-red lamps) so that the adhesive can fill all crevices. This warming up also eliminates wetting of the cold part due to air moisture condensation.

Inorganic fillers are usually added to the adhesives, e.g. quartz flour, graphite, metal powders, etc. They not only reduce material costs, but also make it possible to obtain the same metallic appearance as the part in many cases. The fillers influence the properties of the adhesives, too. For example, the use of graphite or molybdenum sulphide is recommended when repairing friction surfaces.

If fillers are used, it is advisable first to coat the whole surface of the void with a thin resin–hardener mixture; this penetrates into all recesses and crevices of the blow-hole. This thin boundary layer ensures that no air is trapped in the deep branchings of the blow-hole, since the capillary-active resin–hardener mixture allows the air bubbles to pass through the layer. If the blow-holes are flat and do not branch out this boundary layer is not needed.

The adhesive is usually applied over the defective area with a spatula, but applicator guns or, for smaller parts, plastics hypodermic syringes may also be used.

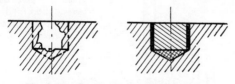

Fig. 8.46 Casting defect rectifying by bonding an inserted plug.

In certain cases, casting defects may be rectified by boring the blow-hole and bonding an inserted plug (Figure 8.46); the fit can be coarse (clearance 0·1 to 0·3 mm).

With porous castings it should be assessed in each individual case whether impregnation is expedient or not. This assessment is necessary since adhesives should be resistant to the temperatures and pressures to which the parts are exposed in service. There are different impregnation techniques, each depending on the size of the castings and the pores. The simplest method is to apply the adhesive over the porous area with a

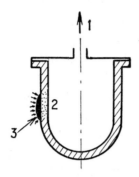

Fig. 8.47 Vacuum impregnation method: 1 – vacuum; 2 – porous area; 3 – adhesive.

brush or a spatula and leave it to penetrate into the pores. However, this is only possible if the pores are large enough. If the pores are fine, the adhesive can infiltrate them only if forced under pressure, using the techniques illustrated in Figures 8.47 to 8.50.

If single or large parts are impregnated, most frequently room-temperature-curing adhesives are used. The casting should be closed and

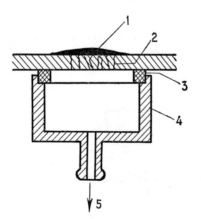

Fig. 8.48 Vacuum cup impregnation method: 1 – adhesive; 2 – pores; 3 – rubber gasket; 4 – cup; 5 – vacuum.

connected to a vacuum pump. The adhesive is applied on the outside over the porous area and is sucked into the pores (Figure 8.47). This method may be modified by using a vacuum cup, as seen in Figure 8.48.

APPLICATION OF ADHESIVE BONDING   155

The cup, made of sheet metal, is connected to a vacuum by means of a flexible hose. The edge of the cup is fitted with a soft rubber gasket.

In the case of small-in-size and large-in-number castings it is better to use liquid heat-curing adhesives because of their longer working life—they will not cure until heat is applied. The castings are filled with

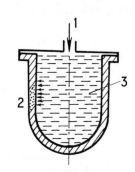

Fig. 8.49 Internal pressure impregnation method 1 — pressure; 2 — porous area; 3 — adhesive.

adhesive, they are closed and under the action of compressed air a pressure, higher than the atmospheric one, is produced. This pressure forces the adhesive into the pores from within (Figures 8.49). Thereafter the surplus adhesive is poured out for re-use, and the impregnated castings are heated according to cure requirements.

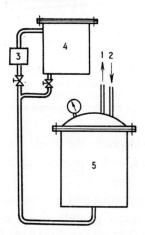

Fig. 8.50 Vacuum-pressure impregnation method: 1 – vacuum; 2 – pressure; 3 – pump; 4 – adhesive container; 5 – hermetic vessel in which the castings are placed.

In the case of batch production, the two methods may be combined by employing an impregnation equipment as shown diagrammatically in Figure 8.50. The castings are placed in a vacuum-pressure vessel where a gauge and an adhesive feeder line are attached. The vessel is closed

L*

hermetically, then evacuated and the proper quantity of deaerated epoxide adhesive is forced in. The vacuum is then relieved and the vessel pressurized. Again liquid heat-curing adhesives are used.

The impregnation of castings was originally used in foundries only as a salvage process for reducing rejection rate. But at present it is acquiring a major role as a processing technique in industry. Impregnation is now

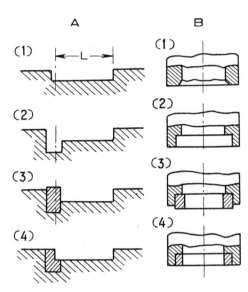

Fig. 8.51 Reconditioning of rejected and worn parts:

(A)
(1) L = prescribed dimension
(2) Remedial groove
(3) Bonded fitting piece
(4) Finished machined workpiece

(B)
(1) Worn part
(2) Boring
(3) Bonded ring
(4) Reconditioned part.

often considered in the original design of pressure-tight parts and parts to be plated, and impregnating equipment and processes are included in production lines.

For repair work and reconditioning of worn parts or scored surfaces, epoxide paste-type adhesives, containing metal powder fillers, are chiefly used. Their consistency before curing and hardness and machinability properties after curing, make them an excellent material for this purpose. Figures 8.51(A) illustrates how it is possible to recondition workpieces rejected because of exceeding the prescribed dimension during machining: a groove is cut in which a flat steel rod is bonded, and then the workpiece is machined to finish dimension. The same technique [42] may be used in reconditioning worn parts, as seen in Figure 8.51(B).

Mention of the sealing of containers and pipes should be made. Leaking areas may be sealed by bonding patches and/or wrapping. Metal sheets may be used as patches, but better results are obtained if the patches and the wrapping are of glass or cotton fabric impregnated with epoxide or polyester adhesive. Containers should be sealed from the side on which pressure is acting. Pipes can easily be sealed on the outside, since the patch is secured by the wrapping.

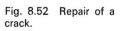
Fig. 8.52 Repair of a crack.

The repair of automobiles, and cracks in blocks and cylinder heads of internal combustion engines is also easily done with epoxide adhesives. The upper part of the crack must first be widened with a chisel. The area is then cleaned and degreased, and heated to 70–80°C. The groove is filled with epoxide paste and covered with a patch; the latter should be rolled with a rubber roller to ensure that no air is trapped between surface and film (Figure 8.52). It is desirable to apply several patches one over the other subsequently.

Broken parts may be reconditioned by bonding only if they are subjected to light loads in service, or if the damage does not influence their function.

9
Bonded Sandwich Structures

For large area structural members designed to resist bending or column loading either thick plates, or thin sheets stiffened with sections or ribs (Figure 7.12) are used in order to obtain members with an adequate rigidity. But in these cases the strength- and rigidity-to-weight ratio is very low.

This disadvantage is avoided with the aid of sandwich composite materials, extensively used in structures in aircraft and space vehicles, and now coming into use increasingly in general engineering. One of the

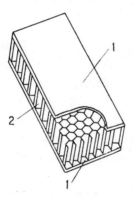

Fig. 9.1 Honeycomb-cored sandwich panel:
1 – face sheets;
2 – core.

most familiar of these composite materials has a honeycomb base. Sandwich structures consist of two thin facings of high-strength material separated by a relatively thick light-weight core (Figure 9.1). The facings serve as load-bearing members. The core is designed to carry shear and loads normal to the facings. It is of primary importance that the core should be rigidly attached to the facings. This cannot be economically effected at the present by any means other than bonding. Consequently, with proper selection of facings, adhesive and core, sandwich composite materials enable high strength- and rigidity-to-weight ratios to be obtained. In addition, vibration damping, and thermal and acoustical insulating properties are achieved. These properties can be varied to suit a wide range of requirements by changing the face and core materials or the design of the core.

Fig. 9.2 Loaded panels: load $F = 16$ kN, breadth = 300 mm, $L = 610$ mm, deflection = 1·5 mm.

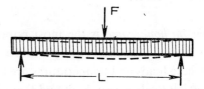

How honeycomb sandwich weight compares with other load-bearing structures is demonstrated in comparing panels loaded as shown in Figure 9.2 and Table 9.1 [1]. It is seen that honeycomb sandwich is by far the lightest of any usable structures.

TABLE 9.1

Structure	Weight (kg)
Honeycomb sandwich	3·5
Nested **I**-beams	5
Steel angles	12
Aluminium plate	15·5
Steel plate	31
Glass-reinforced plastics laminate	38

9.1 Materials for sandwich structures

Facings of sandwich panels are designed to carry the loads acting on the structure. Therefore they are usually made of high-strength materials, such as carbon steel sheets in thicknesses of 0·75 to 1·25 mm, aluminium alloys in thicknesses of 0·25 to 2·5 mm, stainless steel, porcelain enamelled steel and reinforced plastics. In lightly loaded structures, hardboard or plywood can be used for facings.

The core may consist of corrugated sheet, low-density wood or foamed plastics, but the greatest success—measured in terms of strength for a given weight—has been achieved using honeycomb cores. The latter can be made of specially treated paper, aluminium, reinforced plastics, or steel. Their weight is very low—up to 97 per cent of their volume is air and 3 per cent material [2].

Paper honeycomb (usually resin-impregnated kraft paper) offers low cost. Its resistance to compressive loads meets normal requirements. However, since its shear strength is low it can absorb and transmit only limited shear loads. Therefore, additional stiffening members are needed if shear loads are present in sandwich structures with paper honeycomb cores.

Aluminium honeycomb is the most popular core material for sandwich constructions with high strength and minimum weight. It is generally used where loads and temperatures are too high for paper. Its resistance to shearing loads is very high and there is usually no need for stiffening members.

Reinforced plastics honeycomb is more expensive than impregnated paper and aluminium honeycomb. Plastics honeycomb core sandwich panels are mainly used in the aircraft and space industry, and where insulation and dielectric properties are required, e.g. radomes.

Stainless steel and special alloy honeycomb cores are used where end-service and temperature requirements are very high. They are very expensive and, again, are mainly used in aircraft and space vehicles.

Facing and core materials are selected on the basis of strength and environmental requirements and cost. Table 9.2 gives an indication of costs, densities, specific strengths, and temperature limits of some honeycomb core materials.

TABLE 9.2

Type of honeycomb	Approximate cost (£/m^3)	Density (kg/m^3)	Specific compression strength $\left(\dfrac{10^5 \text{ N/m}^2}{\text{kg/m}^3}\right)$	Specific shear strength $\left(\dfrac{10^5 \text{ N/m}^2}{\text{kg/m}^3}\right)$	Temperature limit (°C)
Impregnated paper	40	20 – 120	0·2 – 0·6	0·1 – 0·2	100
Aluminium	120	30 – 130	0·2 – 0·7	0·1 – 0·4	150
Reinforced plastics	500	15 – 150	0·3 – 1·2	0·1 – 0·5	250
Stainless steel	3500 – 7000	90 – 170	0·3 – 1·6	0·1 – 0·6	400 – 500

By varying the cell size, the cell wall thickness, and the material used, various types of honeycomb core with different densities and mechanical properties are obtained. Figure 9.3 shows in principle the approximately proportional relationship between the mechanical properties and the density of the material. It may be seen that it is, in general, possible to

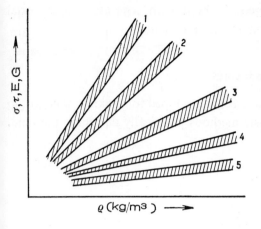

Fig. 9.3 Influence of core density (ρ) on the mechanical properties (σ, τ, E, G) of different core materials: 1 – steel honeycomb; 2 – aluminium honeycomb; 3 – glass-reinforced plastics honeycomb; 4 – impregnated paper honeycomb; 5 – various foams.

achieve with a 'lower rated' core the strength properties of 'higher rated' cores by changing to a higher density.

Honeycomb cores have directional properties. Both shear strength and modulus are higher when the core direction (node bonds) runs parallel to the maximum stress; this results in higher strength. Figure 9.4 shows how the core direction can influence the strength of a beam. Thus, when using honeycomb, load orientation should be considered.

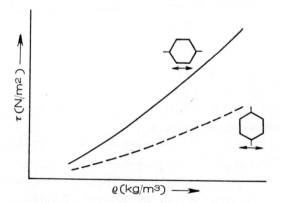

Fig. 9.4 Influence of core direction on the shear strength (τ) of a beam; ρ – core density.

The facing sheets and type of honeycomb are chosen according to the loading expected, though this may not necessarily be the same over the entire structure. For example, the gangway of a passenger aircraft is subjected to greater loads than the flooring beneath the seats. This part can, therefore, be advantageously built with separate denser cored panels, or with different cores in the same panels.

9.2 Manufacture of honeycomb cores

Several patented processes have been developed for making honeycomb cores, but at present two basic methods practically dominate the field.

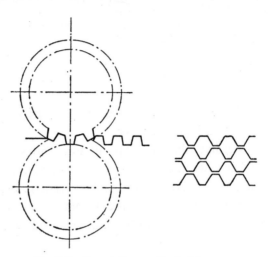

Fig. 9.5 Corrugation method of honeycomb manufacture.

The first method is by forming sheet material between rollers with surfaces grooved in the shape of half hexagon (Figure 9.5). The corrugated sheets are then assembled into large blocks by bonding the mating surfaces (node points) of the cells.

The second basic method is by applying continuous lines of adhesive to sheets of material, building up the sheets into a flat pack, the adhesive lines being in staggered order. The adhesive is then cured under heat and pressure. After curing, the bonded assembly is subsequently cut into strips to the required thickness and expanded by pulling out to form a honeycomb block (Figure 9.6).

Honeycomb blocks are cut on bandsaws into pieces of the required thickness. Stress-free, burr-free and narrow cuts can be made using a 'quenched arc' generated by a band electrode and suppressed by a sprayed coolant. A newer method of cutting metal honeycomb blocks accurately makes use of a needle-like tool. This tool, acting as a conductor for an electrostatic discharge source, heats the metal surface, and oxygen flowing through the tool's hollow needle point does the cutting.

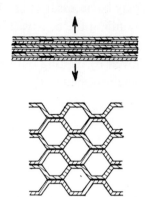

Fig. 9.6 Expansion method of honeycomb manufacture.

The two core-making methods described have their advantages and shortcomings, and defined limits of application, respectively. Cores made by corrugation usually cost up to twice as much, their unit volume is higher and their strength lower because the adhesive usually cures without pressure. In addition, they occupy much storage area as semi-finished products. Expanded cores possess a uniform structure. But this method is inapplicable with thick foils and high-strength materials.

Honeycomb cores are manufactured normally in cell sizes of 3 to 40 mm (these measurements refer to the diameter of the circle inscribed in the cell). They are supplied in slabs up to 2,500 mm long, 1,250 mm wide and 3·5 to 600 mm thick. Such slabs may be joined at their edges to produce honeycomb layers of any required dimension.

9.3 Making up honeycomb sandwich structures

To make a sandwich structure, a piece of honeycomb cut to appropriate size is bonded between the facing sheets. The sandwich structure may be of the simplest form, having flat facing sheets parallel to one another, or it may be of more complicated shape, such as an aerofoil.

If the honeycomb blocks are to be shaped, this can be done by conventional machining techniques. If very close tolerances are required the cores may be sanded on special machines, or they may be machined with 'valve stem' cutters at speeds between 15,000 and 20,000 rev/min. This type of cutter can be made from automobile engine valves, the edges of which have been hardened or are stellite tipped (Figure 9.7). In some instances it may be necessary to stabilize the cores during the machining operations by filling the cells with paraffin wax, or water that is subsequently frozen. After shaping the core to the desired curvature the wax or ice is removed by melting out and the core degreased [3]. The facing sheets are stretch-formed to the same curvature, and the components are bonded together.

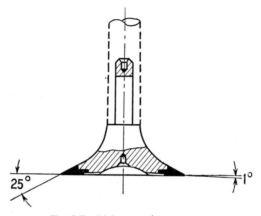

Fig. 9.7 'Valve stem' cutter.

Honeycomb sandwich parts may be produced with two- or three-dimensional curvature. The facing sheets and the core must be formed to shape individually before bonding. The honeycomb can be bent to a radius of about ten times its thickness.

Explosive forming is another method that can be used to produce three-dimensional curved parts from honeycomb sandwich [4]. The most important advantage of this method is the elimination of matching errors of contours, since both facing sheets and core can be formed simultaneously, in either bonded or unbonded condition.

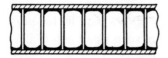

Fig. 9.8 Cross-section through a honeycomb sandwich panel.

Because the edge areas of honeycomb foil are extremely small as compared with the total surface, it is important to create surfaces capable of absorbing the shear loads when the facings are loaded in tension or the core in shear. To achieve this, the configuration of core-to-skin bonds should be as shown in Figure 9.8. It is desirable that the adhesive climbs up along the cell walls to form a fillet joint. This joint configuration can be obtained by the use of adequate quantities of adhesives possessing such wetting and flow characteristics that the fillet is formed during the cure.

Usually, structural adhesives based on thermosetting synthetic resins are used: phenolic, epoxides, or filled phenolic-epoxide blends. Thermoplastic or elastomeric adhesives are used only for low-cost lightly loaded parts.

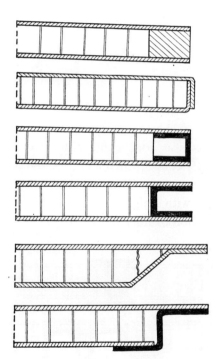

Fig. 9.9 Typical edge treatments for sandwich panels.

9.4 Design features

Normally, it is not required to close off the edges of sandwich panels since most core materials are resistant to normal atmospheric exposure. But this is necessary to prevent accidental damage to the core, provide edge reinforcement and increase the load-carrying capacity [5]. Figure 9.9 shows in principle some typical types of edge treatment.

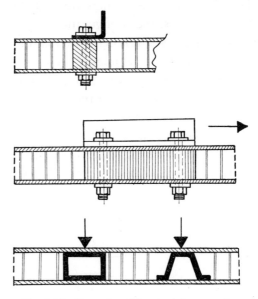

Fig. 9.10   Examples of local reinforcements.

Because honeycomb cores are highly susceptible to crushing from locally concentrated loadings, the core should be reinforced in the load-carrying area. This is achieved by introducing in the attaching areas cores with a higher density, bonding of reinforcing doubler strips, or the insertion of filler blocks (Figure 9.10).

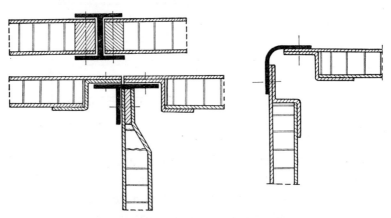

Fig. 9.11   Examples of the type of joints used in honeycomb panels.

BONDED SANDWICH STRUCTURES

The attachment of the panels to supporting members and/or among them can be by bolts through the edging members. In this case a heavier gauge extrusion may be chosen as an edging member. To give an idea on the type of joints in honeycomb structures, a number of examples are shown in Figure 9.11.

9.5 Application of honeycomb sandwich structures

Structural sandwich is most widely used for making aircraft structures. This form of construction enables aircraft components to be simplified, with such members as ribs and frames almost eliminated. The reduction in the number of parts results in significant savings in weight, easier production and less associated paper-work.

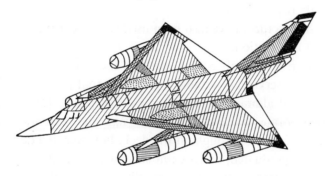

Fig. 9.12 Convair supersonic B-58 bomber; shaded areas show surfaces of bonded panel construction of different types.

For example, in constructing the Convair supersonic B-58 Hustler bomber over 400 kg of adhesives are used. This adhesive took the place of 500,000 rivets. A very high percentage of the surface is bonded sandwich construction (over 230 panels), and as shown in Figure 9.12, it is used on the wings, control surfaces, fuselage, nacelles and exterior doors. In fact, it is used in all areas except radomes and areas where heat requirements call for the use of a steel structure.

Helicopters also make extensive use of sandwich construction. For example, the airframe of the powerful twin-turbine Ch-47 Chinook includes some 75 bonded honeycomb components, of which the major ones are the pod assemblies, the floor frames, the maintenance walkway and the rear loading ramp. An excellent sample of cost saving is shown in the exceptional life of helicopter rotor blades. The service life of a blade when of riveted construction was 90 hours' flying time; the service life of a bonded blade with honeycomb core exceeds 1,200 hours.

Honeycomb panels find extensive use in missiles and space vehicles. For example, in the satellites Ariel-3 and ESRO-2, designed and manufactured in the United Kingdom, epoxide adhesives were used in many items of the equipment [6]. The solar cells, which provide power for the equipment carried, are mounted on bonded aluminium honeycomb panels. Bonded floor panels of the same type are used for mounting the electronic equipment and measuring instruments. The entire octagonal structure of the High Eccentric Orbiting Satellite HEOS-A (launched 5 December, 1968) is made of aluminium honeycomb panels. The cells around the inserts for distributing loads created by fasteners are filled with epoxide resin.

Sandwich structures have also been used with great success in several non-aircraft applications.

The 160-ton hovercraft SRN4, used in the Cross-Channel car and passenger ferry service, is notable for the fact that its decks consist entirely of bonded aluminium honeycomb-cored panels [7]. Some of the panels incorporate honeycomb which has the exceptionally high density of 173 kg/m^3. Throughout the deck, however, the density of the honeycomb is graduated, and its cell structure orientated according to the anticipated stresses. In some cases, different densities are used even within the same panel. In the case of the car deck the aluminium surface is covered with plywood boards which have been given a non-skid surface by the application of bauxite chippings bonded to the plywood by means of an epoxide adhesive. The honeycomb decking in this hovercraft consists of four hundred panels with a total area of 1,350 m^2. The lift fans of some hovercraft are built onto a sandwich base-plate to ensure rigidity without excessive weight.

The combined light weight, rigidity and buoyancy of honeycomb cored panels make them an attractive proposition for boat-building. The structure of hulls made from such panels is sufficiently rigid to require no internal framing. The savings in weight may attain 40 per cent. Furthermore, if the outer skin sustains damage, the non-interconnecting cells honeycomb will isolate the damaged area and the hull will remain watertight.

Adhesive bonded honeycomb sandwich construction is used in the manufacture of large parabolic aerial structures [8]. The units produced weigh approximately 35 per cent less than conventional truss-type structures and have increased rigidity. The increase in rigidity makes it possible to provide significantly higher natural frequencies, while the weight reduction in the sandwich type reflector permits the use of lighter support structures and operating mechanisms. Some of the most noteworthy aerial structures produced are the 18-m diameter microwave reflector aerials used by the Massachusetts Institute of Technology for space tracking purposes. These aerials utilize more than 1,000 m^2 of bonded sandwich construction in the reflectors and other areas.

So far, the largest single application for aluminium honeycomb in the USA is in mats for portable runways for military purposes. Each runway has about 100,000 m^2 of easy-to-handle 1·2 × 1·2 m interlocking honeycomb sandwich panels. Minimum soil preparation is needed. The runway can easily be taken apart and relocated elsewhere.

Where rigidity is needed in machinery, structures are often made of cast iron or heavy forged or welded parts. If these parts are required to move rapidly, a large amount of power is necessary to drive them. In a number of industrial machines such heavy parts have been replaced with honeycomb structures. This has resulted in savings in weight and power, and substantially increased performance if the parts are movable.

An example of the use of honeycomb materials in machinery structures is the body of a pressure butt-welding machine for polyethylene plates with a breadth of 2,000 mm, shown in Figure 9.13 [9]. This machine, developed in East Germany, is built up of impregnated-paper-cored honeycomb sandwich members that provide high torsional stiffness. The facings, edges and joining members are of 2-mm thick aluminium-alloy sheet. The adhesive used for core-to-skin bonding is a glass-mat-supported polyester film. The total weight of the machine is only about 45 kg; this is considerably less than the weight of similar cast or welded structures.

Exact evenness is a primary requirement for doors. This is achieved in honeycomb steel doors, where a paper core is bonded to the two steel facings. Bonding keeps the facings perfectly flat so that there is no buckling. Because of the absence of welded internal stiffeners, surface marring and heat distortion are eliminated. Tests have shown that, in spite of their simple construction, these doors are very sturdy with high resistance to flexing and impact. Honeycomb cores not only serve as insulation but also deaden noise and eliminate the hollow ringing sound typical of the usual metal doors.

Honeycomb panels are especially convenient where a saving in weight is important. This property is applied successfully in the structure of lorry and bus bodies, trailers, for floors, bulkheads and doors in road vehicles, ships and boats. They are very appropriate for light-weight easy-to-handle shipping containers, materials handling pallets and platforms.

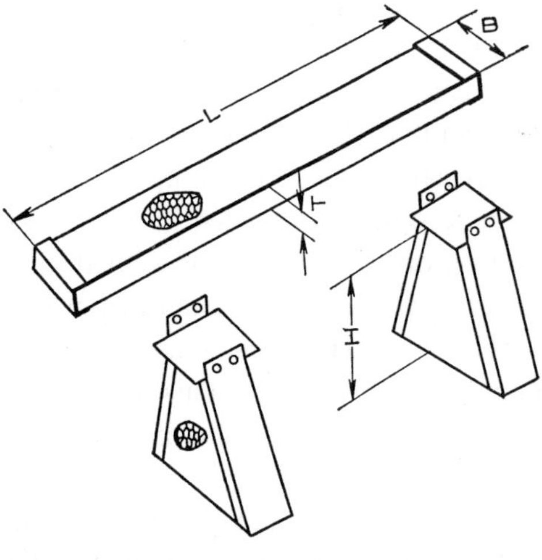

Fig. 9.13 Built-up of the body of a pressure butt-welding machine for polyethylene plates: $L = 2{,}320$ mm, $B = 380$ mm, $T = 85$ mm, $H = 740$ mm.

In architecture and construction weight-saving honeycomb is used for interior and exterior walls, movable and stationary partitions, roofs and floors. Paper honeycomb has wide acceptance as a light-weight, low-cost core material for furniture items, such as table and desk tops, cabinets, counters, and refrigerators. Steel-faced and paper-cored drawing boards are manufactured, having a final PVC finish on the surface. This almost halves the over-all weight, compared with previous wooden ones, with the additional advantage that small drawings may be held in place by suitable magnets. The structure is perfectly flat and rigid, and there is no danger of subsequent distortion with time.

In the USA paper honeycomb panels are used for highway signs because surface flatness can be maintained to close tolerances, so specular dispersion is kept to a minimum and the signs are more easily read.

Because in failure honeycomb absorbs energy at a constant rate with complete dissipation of energy that would otherwise be released in rebound, it is used to absorb shock from air drops of military equipment, for packaging fragile items, etc. For example, in a jet aircraft tail skid an aluminium honeycomb tube weighing only about 0·5 kg substitutes for a 20-kg complex cushioning mechanism; this skid is designed to take up the shock in case the aircraft touches down tail first. And the American lunar landing modules have cylinders of aluminium honeycomb to soften their landing on the moon.

The energy-absorbing properties of honeycomb are also used for protecting human occupants of high-speed vehicles. Bonded sandwich structures have potential automotive 'safety' uses, such as impact-absorbing dashboards and firewalls. It is interesting to note that the main structural beams of Donald Campbell's Bluebird CN7 consisted of aluminium honeycomb sandwich and a very stiff light structure resulted [10]. There is no doubt that the impact-absorbing properties of this construction were the major factor in saving Campbell's life when the car crashed at a speed of over 400 km/hr during its run in the USA; his last boat did not have a similar structure.

There are also a growing number of non-structural applications for honeycomb [1]. For example, honeycomb cores without facing sheets can be used for directionalising air and fluid flow and converting turbulent flow to laminar. These properties are provided by the uniform and parallel cell orientation. Extremely small edge areas of honeycomb foil result in very low pressure losses when gases or liquids pass through the cells. These properties are used in grilles, registers, to control air flow in applications like home fan coolers or open-front freezers in supermarkets, and many others.

Face-less aluminium honeycomb, used widely in the grilles and registers of shielded electronic equipment, acts as a radio frequency shielding device but allows cooling air and light to pass.

The extremely high ratio of surface area to unit volume provided by honeycomb material offers many possibilities in heat-exchange applications, ranging from room air-conditioners to large industrial cooling towers; copper, aluminium and stainless steel are the material used.

Appendix 1
Suppliers and Manufacturers of Specific Adhesives

The table overleaf is a quick guide to manufacturers of adhesives for specific adherend materials and intended to show availability of adhesives for a given material. Note that it is the company's range of products, and not individual adhesives, that is represented here. Detailed advice and data on performance of adhesives in specific cases should be sought from the makers or suppliers, but at some stage most industries will have to carry out their own bonding trials.

Many manufacturers are prepared to produce special-purpose adhesives where an existing range is not adequate for complex adherends and conditions.

Recommended by makers as suitable for bonding:	Ferrous metals	Non-ferrous metals	Glass, Ceramics	Timbers	Natural rubber	Synthetic rubber	Acetals	Acrylics	Aminos	Epoxides	Fluorocarbons	Phenolics	Polyamides	Polyesters	Polypropylenes	Polyethylenes	Polyurethanes	Expanded materials (others)
Adhesive Solutions Ltd.	+	+	+	+	+	+		+	+			+		+	+	+	+	+
Anchor Chemical Co. Ltd.	+	+			+	+		+										+
Arabol-Edwardson Adhesives Ltd	+	+	+		+	+												+
Associated Adhesives Ltd.	+	+	+	+														
Beck, Koller & Co. Ltd.	+	+	+	+			+			+								
Borden Chemical Co. (UK) Ltd.	+	+	+	+	+	+				+		+	+	+	+		+	+
Bostik Ltd.	+	+	+	+	+	+						+	+	+			+	+
BTR Industries Ltd.	+		+	+	+							+						
BXL Plastics Materials Group Ltd.	+	+	+	+	+	+						+	+	+	+			+
Carlton Brown & Ptrs. Ltd.	+	+	+	+	+	+	+	+	+	+	+	+	+	+	+		+	
Casein Industries Ltd.	+	+	+	+	+	+												
CIBA (ARL) Ltd.	+	+	+							+	+							
CIBA Bonded Structures Div.	+	+	+	+						+	+	+	+					
Cornelius Chemical Co. Ltd.	+	+	+	+			+	+	+						+	+		
Cray Valley Products Ltd.	+	+	+	+														
E. M. Cromwell & Co. Ltd.	+	+		+	+	+												
Cyanamid Int. Corp.	+	+	+	+	+	+	+	+	+	+		+	+	+			+	
Devcon Ltd.	+	+	+															
Dow Corning Int. Ltd.	+	+	+	+	+	+		+	+	+		+	+	+		+	+	
Dunlop Chemical Products Div.	+	+	+	+	+	+	+	+		+		+		+			+	+
Durham Raw Materials Ltd.	+	+	+	+	+	+						+	+				+	
Eastman Chemical Int.	+	+	+		+	+	+			+		+	+	+			+	

Company													
Emerson & Cuming (UK) Ltd.	+	+	+	+	+	+	+	+	+	+	+	+	+
Expandite Ltd.	+	+	+	+	+	+	+	+	+	+	+	+	+
Fortafix Ltd.	+	+	+										
Goodyear Tyre & Rubber Co. Ltd.	+	+	+	+	+	+	+	+	+	+	+	+	+
Independent Adhesives & Chemicals Ltd.	+			+	+								
Douglas Kane (Group) Ltd.	+	+	+	+	+		+	+	+	+	+	+	+
Larkhill Rubber Co. Ltd.	+	+	+	+	+	+	+	+	+	+	+	+	+
London Adhesive Co. Ltd.	+	+	+	+	+	+	+	+	+	+	+	+	+
Midland Silicone Ltd.	+	+	+	+	+	+	+	+	+	+	+	+	+
3M Company Ltd.	+	+	+	+	+	+	+	+	+	+	+	+	+
Pertec Ltd.	+	+	+	+									
Pittsburgh-Midland Ltd.	+	+	+	+	+	+	+	+	+	+	+	+	+
Planax Binding Systems Ltd.	+	+	+	+									
Plus Products Ltd.	+	+	+	+	+				+				
Rawlplug Co. Ltd.	+	+	+										
Romac Industries Ltd.	+	+		+			+						
Shell International Chemical Co. Ltd.	+	+	+	+	+	+	+	+	+	+	+	+	+
Sheppy Glue & Chemical Works Ltd.	+	+	+		+								
Unibond Ltd.	+	+	+	+		+							
Unimatic Engineers Ltd.	+	+											
Union Glue & Gelatine Co. Ltd.	+	+	+	+	+	+	+	+	+	+	+	+	+
Witco Chemical Co. Ltd.	+	+	+	+	+	+	+	+	+	+	+	+	+

NOTE: The table above is based on a table published in 'A survey of structural adhesives', *Engng Mater. Des.*, V11(6), pp 946-7 (1968).

Appendix 2

Testing Adhesives for Metals and Adhesive Bonds: Methods, Standards and Specifications

US Standards

Method	Fed. Test Method Standard No. 175 Method No.	ASTM No.
Tensile properties of adhesives	1011·1	D 897–49
Shear strength properties of adhesives by flexural loading	1021	D 1184–55
Shear strength properties of adhesives determined with single-lap constructions by tension loading	1033·1–T	D 1002–53T
Peel of stripping strength of adhesives	1041·1	D 903–49
Peel strength of adhesives (Climbing drum apparatus)	1042–T	D 1781–60T
T–Peel strength of adhesives	—	D 1876–61T
Impact value of adhesives	1051·1	D 950–54
Fatigue strength of adhesives	1061	—
Cleavage strength of metal-to-metal adhesives	1071–T	D 1062–51
Resistance of adhesive bonds to chemical reagents	2011·1	D 896–51
Resistance of adhesive bonds to water	2031	—
Determining the effect of moisture and temperature on adhesive bonds	2052–T	D 1151–56T

US Military Specifications

MIL–A–14042A (Ord) :	Adhesive epoxy
MIL–A–9067B :	Adhesive bonded metal; process and inspection requirements
MIL–A–25463 (ASG) :	Adhesive, metallic structural sandwich construction
MIL–A–8623A :	Adhesive, epoxy resin, metal-to-metal structural bonding
MIL–A–5090D :	Adhesives, heat-resistant, airframe structural, metal-to-metal
MIL–A–13883A (Ord) :	Adhesive, rubber resin, synthetic
MIL–A–1154B :	Adhesive, vulcanized synthetic rubber to metal and to vulcanized synthetic rubber bonding
MIL–A–928A :	Adhesive, metal to wood, structural

APPENDIX II

W. German Standards

DIN 53281, Pt. 1:	Test pieces; Preparation of surfaces to be joined
DIN 53281, Pt. 2:	Test pieces; Manufacturing
DIN 53281, Pt. 3:	Test pieces; Characteristic data of the joining process
DIN 53282:	T–Peel test
DIN 53283:	Tensile shear test on single lap joints
DIN 53284:	Creep rupture test on single lap joints
DIN 53285:	Fatigue test on single shear lap joints
DIN 53286:	Requirements for testing at different temperatures
DIN 53287:	Continuous immersion in liquids
DIN 53288:	Determining the tensile strength perpendicularly to the adhesive plane

References and General Bibliography

References

Chapter 1

1 H. F. HARDRATH, H. A. LEYBOLD, C. B. LANDERS, AND L. W. HAUSCHILD, 'Fatigue-crack propagation in aluminum-alloy box beams', *Ciba Aircraft Bulletin No. 1*.

Chapter 2

1 J. J. BIKERMAN, '*The science of adhesive joints*', Academic Press, New York 1961.
D. KUTSCHA, 'Mechanism of bonding and joint performance', *Adhes. Age*, V11 (11), pp. 37–40 (1968).
2 H. DUNKEN, 'Über physikalische und chemische Adhäsion', *Plaste Kautsch.*, V9(7), pp. 314–317 (1962).
3 A. MATTING AND W. BROCKMANN, 'Adsorption und Adhäsion an Metalloberflächen', *Adhäsion*, No. 8, pp. 343–351 (1968).
4 P. O. SEIDLER, 'Neuere Theorien der Adhäsion von Hochpolymeren', *ibid.*, V7(11), pp. 503–512 (1963).
5 N. A. DE BRUYNE AND R. HOUWINK (EDITORS), '*Adhesion and adhesives*', Elsevier, New York (1951).
6 S. S. VOYUTSKII, '*Autohesion and adhesion of high polymers*', Interscience, New York (1964).
——————, 'Adhesion and autohesion of polymers', *Adhes. Age*, V5(4), pp. 30–36 (1962).
7 V. B. DERYAGIN AND N. A. KROTOVA, '*Adgeziya (Adhesion)*', Academy of Sciences of the USSR, Moscow (1949).
N. A. KROTOVA, '*O skleivanii i prilipanii (On adhesion and sticking)*', Academy of Sciences of the USSR, Moscow (1960).
8 V. VOHRALIK, 'Principles of adhesion', *Rubb. Plast. Wkly*, V140(25), pp. 1006, 1008, 1018 (1961).

Chapter 3

1 British Standard BS 1204: Synthetic resin adhesives.
British Standard BS 1755:Part 1:1967 : Glossary of terms used in the plastics industry.
West German Standard DIN 16921: Adhesives. Processing. Glossary of terms.
2 R. W. CARSON, 'Just to hold parts together assemble with cements', *Prod. Engng*, V37(3), pp. 41–47 (1966).

3 West German Standard DIN 16920: Adhesives. Classification.
4 'Adhesives based on "Epikote" resins', *SHELL Epikote Resins Technical Bulletin.*
5 G. KALISKE, 'Herstellung von Metallklebverbindungen mit Hilfe von Teer-Epoxidharzen', *Plaste Kautsch.*, V12(8), pp. 453–455 (1965).
G. KALISKE AND D. NITSCHE, East German Patent No. 32,032: Klebstoffe aus präpariertem Steinkohlenteer, Steinkohlenrohteer oder anderen Teer-Produkten mit einem Epoxidharz.
6 J. E. HAUCK, 'New high-temperature adhesives', *Mater. Engng*, V65(4), pp. 84–86 (1967).
D. J. CONWAY, 'Metal-to-Metal adhesives for structural application at elevated temperatures', *Adhes. Age*, V11(5), pp. 30–34 (1968).
7 S. LITVAK, 'Polybenzimidazole adhesives', *ibid.*, V11(1), pp. 17–24 (1968); V11(2), pp. 24–28 (1968).
8 J. HERTZ, 'An evaluation of several adhesives in cryogenic applications', *ibid.*, V4(8), pp. 30–37 (1961).
B. PASCUZZI AND J. R. HILL, 'Structural adhesives for cryogenic applications', *ibid.*, V8(3), pp. 19–26 (1965).

Chapter 4

1 S. SEMERDJIEV AND P. PANOV, '*Über das Kleben von Metallteilen mit unreinen Oberflächen*', Paper presented at the International Conference on Adhesive Bonding in Halle (East Ger.), (October 1969).
2 W. A. CARR, R. DUNAETZ AND C. SHINERS, 'Frozen catalyzed epoxy solves instrument bonding problem', *Adhes. Age*, V5(3), pp. 22–25 (1962).
3 P. W. SHERWOOD, 'Microencapsulation: a bid for increased adhesive versatility', *ibid.*, V4(8), pp. 24–26 (1961).
4 'Handling two-component epoxy resin adhesives', *Light Prod. Engng*, V4(7), pp. 27–29 (1966).
5 V. M. HRULEV, USSR Patent No. 140,195: Mixer for cold-setting adhesives.
6 W. GILDE AND S. ALTRICHTER, 'Die optimale Mischung', *ZIS–Mitteilungen*, V10(2), pp. 296–298 (1968).
7 'Double adhesive system improves bonds', *Prod. Engng*, V35(10), p. 50 (1964).
8 H. SCHWARZ AND P. SCHMIDT, East German Patent No. 19,996: Verfahren und Vorrichtung zum Flammspritzen pulverförmiger, pastöser und flüssiger Kunststoffe.
9 V. SUSSMAN, 'Now ... Pre-shaped epoxy adhesives', *Prod. Engng*, V32(21), pp. 43–45 (1961).
S. J. SINER, 'Preformed epoxies can simplify bonding', *Metalwkg Prod.*, V106(5), p. 58 (1962).
10 J. W. JANSON, 'Assembly with adhesives', *Automation*, V10(4), pp. 66–69 (1963).
G. R. NYSTROM, 'Applying adhesives automatically', *ibid.*, V14(12), pp. 60–62 (1967).
11 'Answers to five common adhesive problems', *Mater. Des. Engng*, V63(2), pp. 86–88 (1966).
12 'Adhesive bonding by high frequency heating', *Light Prod. Engng*, V4(7), pp. 21–23 (1966).

13 R. L. HAUSER AND R. E. FISHER, 'Ultrasonics boost strength, speed cure of adhesive', *Mater. Engng*, V67(4), pp. 76–77 (1968).

Chapter 5

1 O. VOLKERSEN, 'Die Nietkraftverteilung in zugbeanspruchten Nietverbindungen mit konstanten Laschenquerschnitten', *Luftfahrtforschung*, V15(1/2), pp. 41–47 (1938).
2 M. GOLLAND AND E. REISSNER, 'The stresses in cemented joints', *J. appl. Mech.*, V11(1), pp. 417–427 (1944).
3 H. WINTER AND H. MECKELBURG, 'Bericht über Metallkleb-Forschungsarbeiten im Institut für Flugzeugbau der DFL, Braunschweig', *Industrie-Anzeiger*, V83(23), pp. 20–27 (1961).
4 K. F. HAHN, 'Lap-shear and creep testing of metal-to-metal adhesive bonds in Germany', Paper presented at the 1959 Symposium on Adhesion and Adhesives in San Francisco, *ASTM Spec. Techn. Publication No. 271* (1961).
5 H. SCHWARZ AND H. SCHLEGEL, '*Festigkeits- und Beständigkeitsuntersuchungen an Metallklebverbindungen*', ZIS, Halle (S.), (1959).
6 'Answers to five common adhesive problems', *Mater. Des. Engng.*, V63(2), pp. 86–88 (1966).
7 H. HERTEL, West German Patent No. 1,204,462: Ausbildung einer Kräfte übertragenden Klebeverbindung von plattenförmigen Bauteilen.
8 I. HLAVAČEK, 'Der Einfluss der Umgebung auf die Festigkeit der mit ChS–Epoxy 1001 geklebten Verbindungen', *Maschinenbau*, V10(9), pp. 372–377 (1961).
9 W. RICHTER, East German Patent No. 18,444: Verfahren für das Verbessern der mechanischen Festigkeit von Kitt-, Leim- oder Klebverbindungen.
R. L. HAUSER AND R. E. FISHER, 'Ultrasonics boost strength, speed cure of adhesive', *Mater. Engng*, V67(4), pp. 76–77 (1968).

Chapter 6

1 B. PASCUZZI AND J. R. HILL, 'Structural adhesives for cryogenic applications', *Adhes. Age*, V8(3), pp. 19–26 (1965).
2 N. S. AKULOV, V. A. KUNAVINA, V. I. AKIMOV AND B. S. OLTCHEV, USSR Patent No. 120,361.
3 C. R. LEMONS, 'Scratching brush sounds out bonding flaws', *Control Engng*, V13(3), p. 108 (1966).
4 R. BOTSCO, 'The Eddy-Sonic test method', *Mater. Evaluation*, V26(2), pp. 21–26 (1968).
5 J. F. MOORE, 'Development of ultrasonic testing techniques for Saturn honeycomb heat shields', *ibid.*, V25(2), pp. 25–32 (1967).
6 W. ALTHOF, 'Zerstörungsfreie Prüfung von Metallklebverbindungen mit Ultraschall-Impuls-Echo-Geräten', *Materialprüfung*, V6(2), pp. 56–62 (1964).
7 S. SEMERDJIEV, 'Zerstörungsfreies Prüfen geklebter Metalle', *Adhäsion*, V10(5), pp. 210–217 (1966).
8 R. J. SCHLIEKELMANN, 'Non-destructive testing of adhesive bonded metal structures', *Adhes. Age*, V7(5), pp. 30–35 (1964), V7(6), pp. 33–37 (1964).
D. F. SMITH AND C. V. CAGLE, 'Ultrasonic testing of adhesive bonds using the Fokker Bond Tester', *Mater. Evaluation*, V24(7), pp. 362–370 (1966).

182 REFERENCES AND GENERAL BIBLIOGRAPHY

9 H. L. DUNEGAN, 'Acoustic emission—a new non-destructive testing tool'. Paper presented at the 1968 ASNT Spring Symposium in Los Angeles.
10 V. N. SHAVYRIN AND L. B. MASEEVA. USSR Patent No. 109,678.
11 C. E. SEARLES, 'Thermal image inspection of adhesive-bonded structure', Paper presented at the 1968 ASNT Spring Symposium in Los Angeles
12 E. W. KUTZSCHER, K. H. ZIMMERMANN AND J. L. BOTKIN, 'Thermal and infrared methods for non-destructive testing of adhesive-bonded structures'. *Mater. Evaluation*, V26(7), pp. 143–148 (1968).
13 S. P. BROWN, '*Detection of flaws in metal honeycomb structures by means of liquid crystals*', Paper presented at the 1968 ASNT Spring Symposium in Los Angeles.
14 C. R. POND, H. H. CHAU AND P. D. TEXERIA, '*Non-destructive testing of honeycomb panels and metal/metal laminates with real-time, stored-beam, holographic interferometry*', Paper presented at the 1968 ASNT Spring Symposium in Los Angeles.

Chapter 7

1 Metallklebverbindungen. Hinweise für Konstruktion und Fertigung. *Richtlinie VDI-2229 (German instruction).*
2 E. A. MAYOROVA AND V. G. AYVAZIYAN, Kleevoe soedinenie tipa val-vtulka pri nagruzkakh krucheniya. *Vestnik mashin.*, V48(7), pp. 55–57 (1968).
3 H. RÜEGSEGGER, 'Nieten—Schweissen oder Kleben?/River-souder ou coller?', *Z. Schweisstech./J. soudure*, V56(12), pp. 463–469 (1966).
4 P. BRENNER, West German Patent No. 1,100,394: Leichtmetallkonstruktionen mit Verstärkungselementen in Form aufgeklebter Blechstreifen zur Aufnahme örtlicher Spannungsspitzen.
5 N. A. DE BRUYNE AND R. HOUWINK, '*Adhesion and adhesives*', Elsevier, New York (1951).
K. FREY, 'Beiträge zur Frage der Bruchfestigkeit kunstharzverklebter Metallverbindungen', *Schweizer Arch. angew. Wiss. Tech.*, V19(2), pp. 33–39 (1953).
H. WINTER AND H. MECKELBURG, 'Zum Entwicklungsstand des Metallklebens', *Metall*, V15(3), pp. 187–199 (1961).
6 H. MÜLLER, 'Statische Untersuchungen an einfach überlappten Leichtmetall-Klebverbindungen', *Fertig.-Tech. Betr.*, V11(1), pp. 40–44 (1961).
7 H. SCHWARZ AND H. SCHLEGEL, '*Metallkleben und glasfaserverstärkte Plaste in der Technik*', pp. 109–110, VEB Verlag Technik, Berlin (Third edition, 1965).
H. SCHLEGEL, 'Vereinfachte Berechnungsmethoden für Metallklebverbindungen, *ZIS-Mitteilungen*, V10(2), pp. 261–271 (1968).
8 H. SCHLEGEL, 'Berechnung von geklebten Rundverbindungen', *Plaste Kautsch.*, V12(8), pp. 469–472 (1965).
9 A. J. DUKE, 'Structural adhesives—to use or to neglect?', *Engng Mater. Des.*, V11 (6), pp. 937–947 (1968).
10 M. J. HILER, 'Adhesive failures . . . Reasons and preventions', *Adhes. Age*, V10(1), pp. 28–31 (1967).
11 G. TRITTLER AND K. DÖRNEN, 'Die vorgespannte Klebverbindung, eine Weiterentwicklung der Verbindungstechnik im Stahlbau', *Der Stahlbau*, V33(9), pp. 257–269 (1964).
12 J. RITCHIE, P. GREGORY, A. J. BANGAY, 'Improvements in bolted joint efficiency by the addition of a cold-setting resin mixture', *Struct. Engr*, V37(6), pp. 175–77 (1959).

REFERENCES AND GENERAL BIBLIOGRAPHY

13 W. SEYFFARTH, East German Utility Model DDR-GM 15,432.
14 K. G. REINHARDT AND C. H. HAENIG, 'Spannungsverteilung in Zugbeanspruchten einschnittigen Punktschweiss- und Punktschweissklebverbindungen', *Schweisstech. (Berlin)*, V17(7), pp. 311-316 (1967).
15 K. G. REINHARDT, 'Zur Wahl von Punktdurchmesser und -abstand bei Punktschweiss-Klebverbindungen', *ZIS-Mitteilungen*, V10(10), pp. 1669-1675 (1968).
16 H. SCHWARZ, 'Punktschweiss-Klebverbindungen', *Plaste Kautsch.*, V12(8), pp. 465-468 (1965).
17 L. KREMPLER, K. MÖSKEN AND K. G. REINHARDT, 'Verfahrensparameter für Stahl- und Aluminium-Punktschweiss-Klebverbindungen bei Einsatz des Kondensator-Impulsschweissens, *ZIS-Mitteilungen*, V9(9), pp. 1253-1264 (1967).
18 K. G. REINHARDT, 'Der Einfluss von Punktschweisslacken, -pasten und -klebern auf die Gefügeausbildung beim Widerstands-Punktschweissen von Stahl', *ibid.*, V9(2), pp. 313-325 (1967).
19 V. N. SHAVYRIN AND L. B. MASEEVA, USSR Patent No. 118,919.
20 W. GILDE, 'Increasing the fatigue strength of butt-welded joints', *Br. Weld. J.*, V7(3), pp. 208-211 (1960).
W. GILDE, H. SCHWARZ AND G. MÜLLER, East German Patent No. 20,891: Erhöhung der Dauerfestigkeit von geschweissten metallischen Bauelementen.
21 T. R. GURNEY, 'Exploratory fatigue tests on plastic coated specimens', *Br. Weld. J.*, V10(10), pp. 530-533 (1963).

Chapter 8

1 H. GROSCH, 'Die Metallklebtechnik, ein wertvolles Verfahren zum Übergang zur Verbundkonstruktion'. *ZIS-Mitteilungen*, V2(1), pp. 55-61 (1960).
2 S. SEMERDJIEV, 'Anwendung der Metallklebtechnik in der bulgarischen Industrie', *ibid.*, V8(4), pp. 570-578 (1966).
3 G. HENNING, East German Patent No. 17,651: Verbundlager, das durch Verkleben mit organischen Bindemitteln hergestellt ist.
4 H. LITZOW, East German Patent No. 35,576: Verfahren zum Verbinden von Maschinenteilen, insbesondere von kegligen Wellen.
5 W. EHRLICH, 'Die Herstellung und der Einsatz von Fräsern mit eingeklebten Schneidplatten', *Feltig.-Tech. Betr.*, V10(8), pp. 459-461 (1960).
6 T. M. RICHARDS AND R. D. WELTZIN, 'Adhesive strengthens riveted assembly', *Adhes. Age*, V8(12), p. 35 (1965).
7 A. MATTING AND K. ULMER, 'Das Metallkleben in der Praxis des In- und Auslandes', *Industrie-Anzeiger*, V85(23), pp. 31-38 (1963).
8 E. F. HESS, 'Structural adhesives for metal parts', *Precis. Metal Mold.*, V21(6), pp. 44-47 (1963).
9 A. MATTING AND G. HENNING, 'Das Kleben von Bauteilen aus Aluminium in der Praxis des In- und Auslandes; *Aluminium*, V42(11), pp. 697-701 (1966).
10 'Epoxy tube plates for heat exchangers', *Ciba Technical Notes* 255.
11 *Ciba Aspekte* (July 1965).
12 F. CROSTER, 'Bonding cooling coils to evaporators of home freezers', *Refrigerator Engng*, V66(3), pp. 48-49 (1958).
13 C. ZELLWEGER, US Patent No. 2,895,633: Receptacles of light metal for liquefied gas.

14 H. BERG, West German Patent No. 1,038,867. Verfahren zum Zusammenhalten von Kolbenringen oder ähnlichen Werkstücken bei der Drehbearbeitung.
15 H. ZACHAU AND S. KLISCHE, 'Einige Beispiele für die Anwendung der Epoxydharze im Strömungsmaschinenbau, *Maschinenbautechnik*, V10(2), pp. 61–65 (1961).
16 W. J. NEUHAUSER AND E. M. TILLER, US Patent No. 3,127,303: Dual-adhesive splicing system.
17 R. A. JOHNSON, 'Bonding of brake liners', *Ciba (ARL) Technical Notes* 203.
   R. E. GREEN, 'Production of brake shoes from strip material', *Machinery, Lond.*, V104 (2687), pp. 1080–1088 (1964).
18 S. B. TWISS, 'How adhesives are used in the modern automobile', *Adhes. Age*, V6(12), pp. 16–22 (1963).
19 'Adhesive bonding bonnet assemblies', *Metalwkg Prod.*, V106(41), pp. 72–73 (1962).
   J. W. JANSON, 'Assembly with adhesives', *Automation*, V10(4), pp. 66 (1963).
20 H. J. KALWEIT AND D. WESSER, 'Technologie und Festigkeitswerte von Metallverklebungen mit dem Klebstoff "Plastaphenal N",' *IfL-Mitteilungen*, V3(8/9), pp. 272–281 (1964).
21 'Adhesive-coated copper simplifies production of die-stamped circuits', *Adhes. Age*, V4(12), pp. 20–22 (1961).
22 I. HLAVAČEK, 'Anwendung des Metallklebens im tschechoslowakischen Maschinenbau, *ZIS-Mitteilungen*, V8(4), pp. 558–569 (1966).
23 M. V. DE JEAN AND M. J. LACY, US Patent No. 3,222,234: Bonding motor parts.
   H. BRECHNA AND A. SCHELLENBERG, Swiss Patents Nos 345,688 and 369,197: Verfahren zur Herstellung eines Blechpaketes.
24 'Laminated busbars', *Electl. Rev., Lond.*, V178(7), pp. 250–251 (1966).
25 'Bonding giant magnetic coils for a synchrotron'. *Adhes. Age*, V7(8), p. 33 (1964).
26 G. TRITTLER, 'Neue Entwicklungen der Verbindungstechnik im Stahlbau', *VDI Z.*, V105(8), pp. 325–331 (1963).
27 J. RITCHIE, P. GREGORY AND A. J. BANGAY, 'Improvements in bolted joint efficiency by the addition of a cold-setting resin mixture', *Struct. Engr*, V37(6), pp. 175–177 (1959).
28 K. DÖRNEN, 'Brücken in Fertigbauweise (Verbundkonstruktionen)', *VDI Z.*, V105(8), pp. 337–346 (1963).
29 G. FRANZ, 'Verbundkonstruktionen aus Aluminium-Legierungen und Stahl', *ZIS-Mitteilungen*, V8(9), pp. 1244–1254 (1966).
30 H. KROSSE, West German Patent No. 1,196,840: Geklebte Stossverbindung.
31 O. JUNGBLUTH, 'Schweiss- und Klebtechnik in der Architektur moderner Stahlhochbauten', *Schweiss. Schneid.*, V12(5), pp. 203–208 (1960).
32 G. E. KLOOTE, 'Metal-faced plywood: light but rigid', *Mater. Des. Engng*, V50(3) pp. 99–101 (1959).
33 'Bonded steel for window frames', *Adhes. Age*, V7(3), p. 33 (1964).
   G. PHILIPP, 'Geklebte Aluminiumfenster', *Aluminium*, V39(2), pp. 121–124 (1963).
   J. AND A. ERBSLÖH, West German Patent No. 1,251,088: Bauelement zur Klebeverbindung von Hohlprofilen, insbesondere zu Rahmen für Fenster, Türen oder dergleichen, Verfahren zu einer derartigen Klebverbindung und Vorrichtungen zu einem solchen Verfahren.

34 J. P. BUSH AND J. A. SCOTT, 'Epoxy adhesive bonds extruded sections of aluminium light pole', *Adhes. Age*, V4(8), p. 24 (1961).
35 M. ROUGIER, French Patent No. 1,371,906.
36 K. GEYS, 'Klebungen im Gleisbau', *Der Bauingenieur*, V39(3), pp. 94–97 (1964).
37 'Araldite in the saving of Abu Simbel', *Ciba Technical Notes* (May 1965).
38 Seven layer bonded sandwich provides skis with permanent camber. *Adhes. Age*, V5(3), p. 33 (1962).
39 D. P. JONES, 'Epoxies for the Tiros satellite', *ibid.*, V4(5), pp. 28–29 (1961).
40 'Redux and Aeroweb in hovercraft', *Ciba Technical Notes* (June 1966).
41 L. P. PAVLOV, '*Kleenye sudovye konstrukzii*', Sudostroenie, Leningrad (1965).
42 H. SCHLEGEL, 'Einsatz von Metallklebstoffen bie der Instandsetzung von Maschinenteilen', *ZIS-Mitteilungen*, V2(10), pp. 684–694 (1960).

Chapter 9
1 *Designing with Hexcel honeycomb*, An advertisement series, Hexcel Products Inc.
2 M. G. BUSCHE, 'Today's honeycomb: light, rigid—more uses', *Mater. Engng*, V66(7), pp. 62–66 (1967).
3 'Use of wax for holding honeycomb for machining', *Machinery, Lond.*, V98(2518), pp. 359–361 (1961).
4 N. N. IDA AND J. T. SNYDER, 'Explosive forming of honeycomb panels', *ibid.*, V103 (2667), pp. 1453–1456 (1963).
5 G. EPSTEIN, 'Adhesive bonds for sandwich constructions', *Adhes. Age*, V6(8), pp. 30–34 (1963).
6 'Ciba products in the Ariel 3 satellite', *Ciba Technical Notes* (November 1967).
7 'Redux and Aeroweb in hovercraft', *ibid.*, (June 1966).
8 R. F. LATCHAW, 'Bonded sandwich construction for parabolic antennas', *Adhes. Age*, V10(7), pp. 24–26 (1967).
9 G. BRINKE AND K. G. REINHARDT, 'Die Stumpf– und Abkantschweissmaschine ZIS–273 für Polyäthylenplatten in Sandwich-Bauweise', *Plaste Kautsch.*, V10(4), pp. 244–248 (1963).
10 'The "Bluebird" CN7', *Ciba (ARL) Technical Notes*, No. 212.

General Bibliography

Books
BECKER, W. E., '*US sandwich panel manufacturing and marketing guide*', Technocomic, Stamford (Conn.), (1967).
BERSUDSKII, V. E., KRYSIN, V. N. AND LESNYH, S. L., '*Proizvodstvo sotovykh konstrukzii (Production of honeycomb structures)*'. Mashinostroenie, Moscow (1966).
BUCHAN, S., '*Rubber to metal bonding*', Crosby Lockwood, London (1959).
GUTTMANN, W. H., '*Concise guide to structural adhesives*', Reinhold, New York (1961).
HENNING, H., KREKELER, K. AND MITTROP, F., '*Untersuchungen über die Kombination Metallkleben-Punktschweissen*', Westdeutscher Verlag, Köln und Opladen (1965).

HENNING, A. H., PEUKERT, H. AND MITTROP, F., *'Auswertung der in- und ausländischen Literatur auf dem Gebiete des Metallklebens'*, Teil II, Westdeutscher Verlag Köln und Opladen (1961).

HOUWINK, R. AND SALOMON, G. (EDITORS), *'Adhesion and adhesives'*, [Volume 1, Adhesives (1965); Volume 2, Applications (1967)], Elsevier, Amsterdam/London.

KANTER, G. G., SHAVYRIN, V. N., ANDREEV, N. H. AND FELDMAN, L. S., *'Kleesvarnye soedineniya v mashinostroenii (Combined welded and bonded joints in mechanical engineering)'*, Tekhnika, Kiev (1964).

KARDASHOV, D. A., *'Sintetitcheskie klei (Synthetic adhesives)'*. Khimiya, Moscow (1968).

KREKELER, K., PEUKERT, H. AND SCHWARZ, O., *'Auswertung der in- und ausländischen Literatur auf dem Gebiete des Metallklebens*, Teil I, Westdeutscher Verlag, Köln und Opladen (1958).

MENGES, G., MITTROP, F. AND DALHOFF, W., *'Auswertung der in- und ausländischen Literatur auf dem Gebiete des Metallklebens*, Teil III, Westdeutscher Verlag, Köln und Opladen (1967).

MIKHALEV, I. I., KOLOBOVA, Z. N. AND BATISAT, V. P., *'Tekhnologiya skleivaniya metallov (Metal bonding technology)'*. Mashinostroenie, Moscow (1965).

MURPHY, J., *'Adhesive bonding: a selected bibliography (1960–1967)'*, Hatfield College of Technology, Hatfield, Hertfordshire.

PARKER, R. S. P. AND TAYLOR, P., *'Adhesion and adhesives'*, Pergamon Press, Oxford (1966).

PERRY, H. A., *'Adhesive bonding of reinforced plastics'*, McGraw-Hill, New York (1959).

RILEY, M. W., *'Plastics tooling'*, Reinhold, New York (1961).

SCHWARZ, H. AND SCHLEGEL, H., *Metallkleben und glasfaserverstärkte Plaste in der Technik*. VEB Verlag Technik, Berlin (1965).

'1968 Adhesives Red Book', Palmerton, New York (1968).

'1968 Book of ASTM Standards', Part 16: Structural sandwich constructions; Wood; Adhesives.

'Adhesives Directory'. A complete Guide to the British Adhesives Industry. A. S. O'Connor & Co., Richmond, Surrey.

'Arbeitsblätter für das Metallkleben', Aluminium-Verlag, Düsseldorf (1962).

'Arbeitsblätter für die Kleb- und Giessharztechnik', Kammer der Technik, Berlin (East).

Surveys in journals

DUKE, A. J., 'Structural adhesives—to use or to neglect?' *Engng Mater. Des.*, V11(6), pp. 937–947 (1968).

POHL, A., 'Klebverbindungen, Metallkleben und Flüssigkunstharzanwendung', *Tech. Rdsch.*, V59(37), pp. 17–19 (1967); V59(50), pp. 57–61 (1967); V59(51), pp. 33–37 (1967); V59(52), pp. 41–45 (1967); V59(54), pp. 41–45 (1967); V60(1), pp. 41–43 (1968); V60(23), pp. 41–45 (1968); V60(24), pp. 9–15 (1968); V60(25), pp. 37–39 (1968).

RIDER, D. K., 'Bonded metal assembly', *Prod. Engng*, V35(11), pp. 85–98 (1964); V35(12), pp. 75–78 (1964).

TWISS, S. M., 'Structural adhesive bonding', *Adhes. Age*, V7(12), pp. 26–31(1964); V8(1), pp. 30–34 (1965).

WICK, C. H., 'Adhesive bonding', *Machinery Prod. Engng*, V112(2892), pp. 732–739 (1968); V112(2894), pp. 836–844 (1968); V112(2896), pp. 932–940 (1968); V112 (2898), pp. 1028–1036 (1968).

Index

'ABS': *see* Elastomeric adhesives
Abu Simbel masonry, 147
Acetone, 31, 35, 49
Accelerators (for adhesives), 13, 23
Acoustic inspection methods, 90–91, 94
Acrylonitrile-butadiene-styrene ('ABS'): *see* Elastomeric adhesives
Activator (for adhesive), 25
Adherends: *see* Surfaces
Adhesion: *see also* Adhesives, *also* Bonds
    actual/theoretical strength, 10–11, 55, 88–89
    and cohesion, difference, 55, 65
    anomalous, despite oil, 33
    fails in high humidity, 73
    hydrodynamics, 16
    instant, 16, 18
    not purely mechanical, 8
    physics of, 7, 28, 55
    various theories, 8–10, 16, 17, 106
Adhesives generally: *see also*
    Adhesion, *also* Curing, *also*
    Dispensing, *also* Economics, *also*
    Elastomeric adhesives, *also*
    'Hybrid' ('alloy') adhesives, *also*
    Thermoplastic adhesives, *also*
    Thermosetting adhesives
    categories/characteristics, 6, 15–19, 55, 73
    ceramic, 26
    combinations of, 44, 76, 128
    conducting, 137
    constituents, 12–13, 18–20
    contact type, 17
    cured, 88
    differential expansion, 51, 56, 75, 97
    domestic, 23
    duplex use, 44, 76, 128
    forms of (film, liquid, rod, etc.), 15, 21–23, 26, 113
    frozen, 38
    future, 6
    growth of demand, viii
    high-temperature, 6, 25–26, 68–69
    household, 23
    low-temperature, 26–27, 68–69
    optimum thickness, 55, 63, 65–66, 100–102
    pioneering uses, 1, 108–110, 129, 149
    pressure sensitive, 16
    salvaging castings, 152–156
    selecting, 14, 21, 173–175
    structural, defined, 19
    suppliers listed, 173–175
    weaken through ageing, 73–74, 87
Aeroplanes: *see* Bonds, *also* Sandwich
'AGE' (Allyl glycidyl ether), 23
Ageing, 73–74, 87
Agitators (in mixing), 39
Aircraft: *see* Aeroplanes
Air cushion vehicles, 150, 168
Allergy: *see* Hazards
'Alloy' adhesives: *see* 'Hybrid' adhesives
Allyl glycidyl ether ('AGE'), 23
Aluminium oxide, as filler, 23
Anaerobic setting: *see* 'Hybrid' adhesives
Ariel-3 satellite, 168
Asbestos:
    as filler, 13, 23
    fabric, as reinforcement, 16
Aswan High Dam, 147
Autoclave, 51
Automation, 47–48

Bathyscaph, 151

'BDMA' (Benzyldimethylamine), 23
Benzene, 35
BF_3 (curing agent), 23
Binders (in adhesives), 12
Bluebird (racing-car), 171
Boeing 727 aeroplane, 108
Bolts:
    combined with bonding, 4, 108–113, 139–141
    obviated, 2, 4
Bonds, practical examples: *see also* Bonds, principles of, *also* Sandwich structures
    and bolts, 109–111, 138–139
    and welds, 112–114
    as reinforcement, 104
    augmenting fastenings, 4, 108–113, 123, 138
    brazing obviated, 2–3, 125–127
    demountable, 123, 124
    duplex adhesives, 44, 76, 128
    for aeroplanes, 1, 4, 149–150, 158
    for architectural components, 142–146
    for automobiles, 128–132
    for brake/clutch linings, 128–129
    for electrical assemblies, 133–138
    for hammer-heads, 123
    for joining castings, 124–125
    for lamp posts, 145–146
    for machine parts, 2, 4, 116–121
    for magnetic assemblies, 124, 132–133, 135, 137
    for marine craft, 150–152, 168
    for masonry-keys, 147
    for patching metal, 157
    for railway track, 146–147
    for skis, 148–149
    for spacecraft, 27, 150, 158, 168
    for steelwork, 140–142, 147
    for studs, 115–116
    for tool-tips, 122–123
    for vehicle bodies/parts, 128–132
    historic, viii, 1, 109–110, 129, 149
    machining saved, 119, 120, 127
    of stiffeners, 103–104

patent-rights, viii
protection of, 74
types of joints, 98–103, 128
Bonds, principles of: *see also* Bonds, practical examples, *also* Test data
    fail at ends, 58, 76
    dimensioning empirical, 105–108
    geometry, 57–60, 63–65, 75–76, 105–108
    grooved, 102, 121
    in sandwich structures, 164–165
    'joint factor', 60, 73
    optimum thickness, 55, 63, 65–66, 100–102
    peeling, 67, 75, 97–98, 103
    protection of, 74
    strengthening-methods, 44, 75–76, 97–98
    strength figures, 18, 55, 75, 77, 101, 107, 109–110, 115
    stress-behaviour, 60–62, 65, 97
    stress-distribution, 56–58, 62, 76, 97
    weakening of, 6, 73–74
Brake linings bonded, 128–129
Bridge incorporating adhesive, 109–110, 139–140

Campbell, Donald, 171
Castings (metal):
    assembled by bonding, 124–125
    cleaning of, 32–33
    defects filled, 153–155
    patched, 157
'Catalyst' gun, 43
Chinook helicopter, 167
Cholesteric crystals, 96
Clamping during bonding:
    criteria, 52
    'fast' adhesives instead of, 24, 44, 52
    jigs/methods, 52–53
    non-stick fluids, 53
    to force mating, 52
Cleaning: *see* Surfaces
Clutch linings bonded, 128–129
Coal tar, as extender, 24
Cohesion defined: *see* Adhesion

INDEX

Cold-setting adhesives: *see* 'Hybrid' adhesives
Colouring, purpose of, 13
Contaminents, influence of, 31–33
Convair B-58 aeroplane, 167
Crack propagation, 3–4, 114
Creep, 69–70
Curing: *see also* Clamping, *also* Processing
  agents, 12, 16, 23
  anaerobic, 17, 24
  chemistry of, 19
  delay tolerable, 46
  electrical, 51
  machining after, 54, 164
  obviated, 16–17, 24
  polymerization, modes of, 19
  pressures, 16, 22
  temperatures, 16, 22, 49–50, 54
  times, 16, 18, 20, 24, 26, 49–50
'Curtain' method of coating, 45
Cyanoacrylate compounds, 24, 44, 53

De Havilland Aircraft Company, 149
Dibutyl phthalate, 23
Dicyandiamide ('DICY'), 23
Dielectric heating, 52
Diethylaminopropylamine, 23
Diethylene triamine ('DTA'), 23
Diluents, reactive, 13
Dispensing of adhesive:
  cheapest-mix calculations, 40–42
  cold-storage, 38
  'curtain' coating-method, 45
  delay before curing, 37, 46
  duplex adhesives, 44
  heat of reaction, 38
  mass effect, 38
  metering equipment, 40, 47–48
  mixture critical, 36, 38, 40
  non-stick vessels, 39
  packaged micro-capsules, 38
  placing as film, 47
  placing as pellets, 46
  placing as powder, 46
  placing as rod, 46
  placing by machine, 45–48
  placing manually, 43–45
  pre-mixed, frozen, 38
  time before use, 37, 46
'DTA' (Diethylene triamine), 23

Economics:
  brazing obviated, 125–127
  bulk-buy savings, 40
  choice of adhesive, 14
  electric curing, 51
  high-temperature premium, 26
  improved by extenders, 24
  labour-saving, data on, 5
  machining avoided, 119–120, 127
  'one-part' adhesives, 25
  proportions in mix, 40–42
  relative, tabulated, 20
  salvaging of parts, 152–157
Egyptian temple masonry, 147
Elastomeric adhesives: *see also* Adhesives generally
  in 'hybrid' formulations, 18–19, 21, 22
  instant stick, 18
  suppliers listed, 173–175
Electrostatic cutting, 162–163
Epoxide compounds: *see* 'Hybrid' adhesives, *also* Thermosetting adhesives
ESRO-2 satellite, 168
Ethylene-glycolacrylate, 25
Examples: *see* Bonds, practical examples
Explosive forming, 164
Extender(s):
  coal tar as, 24
  economy calculations, 40–42
  purpose of, 13

Fibres:
  as reinforcement, 16, 75, 97
  glass, as filler, 13, 39
Fillers (in adhesives):
  economy calculations, 40–42
  materials used, 13, 23, 39
Fire: *see* Hazards
Flame-spraying, 46
Flexibilizers, 13

Fokker F 27 and F 28 aeroplanes, 149–150
Fokker bond-testing instrument, 92
Freon (gas) resisted, 127

Glass fibres: see Fibres
Graphite, as filler, 13, 23

Hardeners, see Curing agents
Hazards:
  fire, 49, 143
  to food, 48
  to personnel, 28, 48–49
Heating methods: see Processing
Helicopter, 167
Hexahydrophthalic anhydride ('HPA'), 23
High Eccentric-Orbiting Satellite (HEOS-A), 168
Holography, 96
Honeycomb cores, see Sandwich structures
Honeycomb, face-less
  as heat exchangers, 171
  as radio shields, 171
  for directing fluids, 171
Hornet aeroplane, 149
Hot-melt: see Thermoplastic adhesives
Hovercraft, 150, 168
'HPA' (Hexahydrophthalic anhydride), 23
Hustler aeroplane, 167
'Hybrid' ('alloy') adhesives: see also Adhesives generally
  cold-setting, 24
  commonest structural type, 18
  data tabulated, 20
  epoxide resist Freon, 127
  favoured for bonds, 21
  high-temperature, 25–26
  'one-part', 17, 24–25
  quick-setting, 17, 24, 44, 53
  suppliers listed, 173–175
Hydrochloric acid, 34

Induction heating, 48, 52
Inhibitors in adhesives, 13
Inspection: see Test

Interferometry, 96
Irradiation: see Radiation

Joints: see Bonds

London-dispersion forces (adhesion theory), 9

Machining:
  after curing, 54
  obviated, 119, 120, 127
  of sandwich structures, 163–164
Massachusetts Institute of Technology, 169
Materials, see Surfaces
Melamine resins: see Thermosetting adhesives
Metal powder as filler, 13, 23, 39
Methanol, 34, 35
Methylene chloride, 31
Mica as filler, 23
M.I.T., 169
Mixing: see Dispensing
'MPD' (m-phenylenediamine), 23

'Neoprene': see Elastomeric adhesives
Nitrile (rubber): see Elastomeric adhesives

'One-part' adhesives: see 'Hybrid' adhesives
Ovens, 51
Oxalic acid, 34

Patent-rights, viii
'PBI' (Polybenzimidazole), 26
Perchloroethylene, 31
Petrol (for cleaning), 31
'PGE' (Phenyl glycidyl ether), 23
Phenolic /-rubber /-vinyl compounds: see 'Hybrid' adhesives, also Thermosetting adhesives
'PI' (Polyimide) compounds: see Thermosetting adhesives
Piccard, Professor, 151
Pigments, purpose of, 13
Plasticizers, 13

INDEX

Polyacrylic/polyacrylate compounds, 17–18, 24–25, 44, 53
Polyamides: *see* Thermoplastic adhesives
Polybenzimidazole ('PBI'), 26
Polychloroprene ('Neoprene'): *see* Elastomeric adhesives
Polyester compounds: *see* 'Hybrid' adhesives, *also* Thermosetting adhesives
Polyethylene: *see* Thermoplastic adhesives
Polyimide ('PI') compounds: *see* Thermosetting adhesives
Polymerizing: *see* Curing
Polymers, aromatic, 26
Polysulphide: *see* Elastomeric adhesives
Polyurethane compounds: *see* 'Hybrid' adhesives, *also* Thermosetting adhesives
Polvinyl /-acetate /-formal, 23
Potassium dichromate, 35
Precautions: *see* Hazards
Processing: *see also* Clamping, *also* Curing, *also* Surfaces
  differential expansion, 51, 54
  double-adhesive bond, 44, 53
  heating methods, 48, 51–52
  'hot-shot' (rapid), 52–53
  machining hints, 54
  selective heating, 51
  slow cooling, 54
  thermocouples used, 53
  ultrasonic heating, 52
  without ovens, 48, 51
Proton synchrotron, 136

Quartz flour, as filler, 13
Quick-setting: *see* 'Hybrid' adhesives

Radiation:
  infra-red, 95
  isotope, for inspection, 94
  strengthens bonds, 77
  ultra-violet, for inspection, 95–96
  X-ray, for inspection, 94

Reactive diluents, 13, 23
'Redux 775' (adhesive), 60
Rivets:
  combined with bonding, 108–113
  obviated, 2, 4, 167
Rubber, synthetic: *see* Elastomeric adhesives

Safety: *see* Hazards
Sandwich structures (honeycombs):
  as shock-absorbers, 171
  core sizes/forming, 162–163
  directional strength, 161
  edge closures, 165, 167
  explosive forming, 164
  fabrication, 164
  for aerial bowls, 169
  for aeroplanes, 158, 160, 167–168, 171
  for architectural components, 170
  for highway signs, 170–171
  for marine craft, 168
  for runways, 169
  for spacecraft, 158, 160, 168, 171
  for vehicle bodies/parts, 169–171
  in machinery, 169
  machining of, 162–164
  materials for, 159–160
  reinforcing of, 166
  strength/temperature limits, viii, 159–161
Satellites, 27, 150, 158, 168
'SBR' (Styrene-butadiene): *see* Elastomeric adhesives
Screening, radio, 171
Silicone:
  as non-stick fluid, 53
  compounds standing 300°C, 25
  elastomeric adhesive, 18–19, 21, 23
Sodium carbonate/dichromate/ hydroxide/metasilicate, 31–35
Solvents:
  defined, 13
  escapement paths, 18, 21, 43
  for stray adhesive, 53
  obviated, 22–24

Spacecraft: *see* Bonds, practical examples, *also* Sandwich structures
Specifications, standard, 78, 176–177
Spray-gun, 43, 46
SRN 4 Hovercraft, 168
Stabilizers (in adhesives), 13
Standard tests, 78, 176–177
'Structural' adhesives: *see* 'Hybrid' adhesives
Styrene-butadiene ('SBR'): *see* Elastomeric adhesives
Sulphuric acid, 34–35
Surfaces:
  anodizing deprecated, 35
  ceramic, preparing, 36
  contamination, tolerable, 33
  degreasing, 31–33
  etching recipes, 34–35
  flame-treatment, 35
  glass, preparing, 36
  immersion-cleaning deprecated, 31
  leather, preparing, 36
  metal-grit blasting, 29, 33
  of castings, cleaning, 32–33
  oxide removal, 34
  physics of, 28
  plastic, preparing, 35
  priming of, 42–43
  protecting after cleaning, 42–43
  roughness criteria, 30–31, 66, 102
  rubber, preparing, 35
  sand-blasting, 29, 33, 35, 122
  ultrasonic cleaning, 33
  wetting-tests, 29
  wood, preparing, 36
Synchrotron, 136

Talc, as filler, 13, 23
Tar, as extender, 24
Temperature limits: *see also* Test
  and bond failure, 68–69, 97–98
  decomposing/softening, 19, 25
  differential expansion, 51, 56, 75, 97
  extended, 25–27
  of phenolic-nitrile, 22
  of sandwich structures, 160

Test data/methods:
  acoustic, 90–91, 94
  age-weakening, 73–74, 87
  capacitance, 94–95
  cholesteric, 96
  corrosion, 88
  creep, 70–71, 84
  drum devices, 82–83
  eddy-current, 91
  environmental, 73–74, 88
  fatigue, 71–73, 84–86, 112–114
  flaw-detection, 90–96
  Fokker instrument, 92
  holographic, 96
  humidity, 73–74, 88
  impact, 73, 86
  impedance, 91–93
  infra-red, 95
  intermodulation, 93–94
  peel, 67–68, 81–83
  penetrating-fluid, 95
  permanence, 87
  preparation for test, 89
  qualification, 89
  resonance, 91–93, 96
  sand-patterns, 96
  specifications/standards, 78, 176–177
  strength, 78–81, 109
  temperature limits, 68–69, 87–88
  thermal image, 95–96
  thermal infra-red, 96
  trepanning, 96
  ultrasonic, 91–93
  vacuum-bell, 95
  void-detection, 90–96
  X-ray, 94
Theory: *see* Adhesion
Thermoplastic adhesives: *see also* Adhesives generally
  characteristics/cost, 18
  hot-melt, 16, 18, 48
  in 'hybrid' formulations, 19
  suppliers, listed, 173–175
Thermosetting adhesives: *see also* Adhesives generally
  characteristics/cost, 19

INDEX

Thermosetting adhesives—*contd.*
   high-temperature compounds, 26
   in 'hybrid' formulations, 19
   resist Freon, 127
   suppliers listed, 173–175
'Thiokol': *see* Elastomeric adhesives
Thixotropic agents, 13
Tiros satellite, 150
Trichloroethylene, 31

Ultrasonic:
   bond-testing, 91–93
   cleaning, 33
   heating-method, 52
   strengthening-effect, 77
Urea resins: *see* Thermosetting
   adhesives

Uses: *see* Bonds

Vacuum:
   in bond-testing, 95
   in clamping, 53
   in salvaging castings, 152–156
Vibrations, 77
Vinyl: *see* Thermoplastic adhesives

Welds combined with bonding, 108–113
Welfare: *see* Hazards
Westland Aircraft, 150
Wetting:
   agents, 13
   and strength, 55, 164–165
   tests for, 29

X-ray: *see* Radiation